An Introduction to CAD for VLSI

An Introduction to CAD for VLSI

by

Stephen M. Trimberger
VLSI Technology, Inc.

Kluwer Academic Publishers
Boston/Dordrecht/Lancaster

Distributors for North America:
Kluwer Academic Publishers
101 Philip Drive
Assinippi Park
Norwell, Massachusetts 02061, USA

Distributors for the UK and Ireland:
Kluwer Academic Publishers
MTP Press Limited
Falcon House, Queen Square
Lancaster LA1 1RN, UNITED KINGDOM

Distributors for all other countries:
Kluwer Academic Publishers Group
Distribution Centre
Post Office Box 322
3300 AH Dordrecht, THE NETHERLANDS

Consulting Editor: Jonathan Allen

Library of Congress Cataloging-in-Publication Data

Trimberger, Stephen, 1955-
  An introduction to CAD for VLSI.

  Includes bibliographies and index.
  1. Integrated circuits—Very large scale
integration—Design and construction—Data processing.
2. Computer-aided design. I. Title.
TK7874.T753   1987   621.381′73′0285   87-2841
ISBN 0-89838-231-9

Copyright © 1987 by Kluwer Academic Publishers

All rights reserved. No part of this publication may be reproduced, stored in a retrieval system, or transmitted in any form or by any means, mechanical, photocopying, recording, or otherwise, without the prior written permission of the publisher, Kluwer Academic Publishers, 101 Philip Drive, Assinippi Park, Norwell, MA 02061.

Printed in the United States of America

to
leon and carol

CONTENTS

Preface	xiii
Chapter 1. Integrated Circuit Design	**1**
Introduction to Integrated Circuit Design	1
The Integrated Circuit Design Process	6
Variances from This Model of the Design Process	11
Managing the Complexity of a Design	12
Exercises	15
References	15
Chapter 2. Parsing	**17**
Reading Files	17
Describing a Data Format	18
An Example	24
Parser Overview	25
Implementation of the Parser	26
Internal Parser Procedures	29
The Token Scanner	31
Building a Data Structure	32
Data Structures	33
The Semantics Module	34

Memory Usage Issues	38
Writing Layout Files	38
Exercises	39
References	41

Chapter 3. Graphics 43

Output Devices	43
Graphical Output Primitives	45
Virtual Screen Coordinates	45
User Coordinates	47
Clipping and Viewports	47
Summary of Graphics Output Features	56
Summary of Coordinate Systems	56
Color	57
Graphical Input	61
Dynamic Displays	62
Exercises	65
References	67

Chapter 4. Plotting Layout 69

The Core of a Plotter	69
The Command Loop	70
Loading a File	71
Displaying a Cell	72
Necessary Additions	73
Desirable Features	74
Practical Considerations	75
Considerations for Pen Plotters	75
Considerations for Raster Plotters	77
Exercises	79
References	80

Chapter 5. Layout Editor — 81

Overview of a Layout Editor	81
Turning the Plotter Into an Editor	82
Pointing As an Alternative to Typing Positions	84
Filters	85
Editor State	85
Modes and Modeless Editors	86
Menus As an Alternative to Typing Commands	88
Data Structures	88
Essential Features	97
Desirable Features	100
User Interface	102
The Database Alternative To Reading and Writing Files	108
Symbolic Layout	109
Schematic Editor	113
Exercises	115
References	117

Chapter 6. Layout Language — 119

Embedded Language	120
A Simple System	120
A More Usable System	121
Layout Language Procedures	122
Data Structures	124
Example	124
Symbolic Layout	125
The Example Again	129
Using Bounding Boxes and Connectors	130
Reading Layout Files	131
Using Language Facilities	131
Parameterized Cells	133

The Layout Language Module As a LayoutParser	135
A Procedural Netlist Generator	136
Drawbacks of a Layout Language	136
Exercises	137
References	138

Chapter 7. Layout Generators — 141

Parameterized Cells	141
PLA Generator	149
Introduction to Silicon Compilation	159
Datapath Compiler	159
Introduction to Placement and Routing	168
Exercises	177
References	179

Chapter 8. Layout Analysis — 183

Overview and Background	183
Design Rules	184
Object-Based DRC	185
Edge-Based Layout Operations	187
Handling Intersecting Edges	191
Polygon Merging	191
Arbitrary Boolean Operations On Layout Layers	198
Resizing Layout	199
Using Bloat and Shrink to Perform a Design Rule Check	202
A More Efficient DRC	203
Connectivity	210
Performance Optimization Considerations	212
Reporting DRC Errors	213
Ambiguous Corner Checks	214
Glitches	216

Technology File		218
Fast Sorting For Edge Files		221
Circuit Extraction		221
Options		227
Mask Tooling		229
Other Algorithms		229
Exercises		231
References		232

Chapter 9. Simulation 235

Types of Simulators		236
A Simple Behavioral Simulator		237
Time and the Event Queue		239
Modelling Shared Busses		241
A Logic Simulator		242
Controlling and Observing the Simulation		246
Managing the Event Queue		249
Simulating Transistors		250
Transistor Simulation Example		261
Debugging Commands		264
Enhancements to the Transistor Simulation Algorithms		264
Static Checks		269
Multi-Level Simulation		272
High-level Input		274
Testing Integrated Circuits		275
Fault Simulation		276
Exercises		278
References		280

Index 283

PREFACE

The last decade has seen an explosion in integrated circuit technology. Improved manufacturing processes have led to ever smaller device sizes. Chips with over a hundred thousand transistors have become common and performance has improved dramatically. Alongside this explosion in manufacturing technology has been a much-less-heralded explosion of design tool capability that has enabled designers to build those large, complex devices. The tools have allowed designers to build chips in less time, reducing the cost and risk. Without the design tools, we would not now be seeing the full benefits of the advanced manufacturing technology.

The Scope of This Book

This book describes the implementation of several tools that are commonly used to design integrated circuits. The tools are the most common ones used for computer aided design and represent the mainstay of design tools in use in the industry today. This book describes proven techniques. It is not a survey of the newest and most exotic design tools, but rather an introduction to the most common, most heavily-used tools. It does not describe how to use computer aided design tools, but rather how to write them. It is a view behind the screen, describing data structures, algorithms and code organization.

This book covers a broad range of design tools for Computer Aided Design (CAD) and Computer Aided Engineering (CAE). The focus of the discussion is on tools for transistor-level physical design and analysis. These tools are applicable across many design disciplines and across all phases of the design process. They comprise a minimum set of tools needed to design working chips consistently.

This book does not describe every tool or type of tool for integrated circuit design. Some topics are omitted because of limited space and time, because they are not sufficiently different from tools that are described in the book, because they are too new and have not proven their worth or because they comprise a large and separate body of knowledge that requires a more thorough treatment than would fit into this book. I have attempted, at least, to mention in context the tools I have omitted to allow the reader to pursue them separately.

This book is the book I wish I could have read when I got started. It introduces tools across the whole spectrum of design tasks. It describes the most successful algorithms and techniques in a form that is comfortable to a novice in the field. I am somewhat disappointed with how little actually fits into this book. Readers should be warned that it is, as it says in the title, an introduction. There is much to computer aided design that is not here, but here is a starting point.

The Intended Audience

This book is intended to tell a programmer how to write software for computer aided design of integrated circuits. It does not discuss theoretical results or describe algorithms in rigorous mathematical terms. It describes computer software and does so in a language that should be comfortable to programmers. As a secondary goal, this book gives the rationale behind decisions made when developing CAD software to give the reader the ability to make these design decisions himself with consideration of his problem. These decisions can make the difference between a tool that is widely useful and one that is a simple experimental toy.

This book assumes a solid knowledge of programming and that the reader has had an introduction to data structures and algorithms. Although no integrated circuit design experience is required, the whole exercise of writing tools for integrated circuit design is pointless without some knowledge of the target domain. A programmer could learn to write design tools without the integrated circuit design expertise, but some knowledge is needed to understand the design tradeoffs.

Although this book is suitable as a textbook for advanced undergraduates, it is more suited to early graduate students in computer science who have the programming experience and have been exposed to integrated circuit design issues. This book is also useful in an industry setting where there is interest in developing or interfacing computer aided design tools. A programmer in such an environment should find the discussion in this book revealing about the software with which he is working.

Current practitioners of CAD programming are often very knowledgeable in their specific domain, but somewhat ignorant of techniques in other areas. This book will round out their breadth of knowledge of CAD tools across the range of integrated circuit design.

Overview of the Book

The book adopts an informal style. Where precision is necessary, it is in the form of pieces of code. Ideas are motivated and developed from simple solutions to more complex ones. Later chapters build on earlier ones, but each chapter is understandable and complete on its own.

The book starts with an overview of the design process and of CAD tools and proceeds to summaries of parsing and interactive graphics techniques. The goals of these introductory chapters are to provide background material and to emphasize the parts of those fields that are applicable to CAD. If the reader is familiar with these fields, he may safely skip the introductory chapters. Integrated circuit design, parsing and computer graphics are fields of study in their own right and the summaries given in this book do not cover those fields in great depth.

Later sections of the book describe graphical techniques, procedural techniques and analysis; each chapter is dedicated to one type of tool. Each introduces the tool and describes its implementation in a developmental way. Simple options are discussed and their drawbacks are used to provide the rationale for the complexity of the final system. These chapters are intended not only to give the reader the knowledge of the methods and techniques of building tools, but the reasons for the complexity of those tools. It is hoped that the reader will be able not only to implement these tools, but to make the decisions necessary to further develop the tools described in this book.

At the end of each chapter I briefly discuss related tools. The assignments at the end of the chapter serve two purposes: first, to focus the reader's thinking to issues that he should consider during the design and second, to give the reader experience developing CAD code in a reasonable framework. Every chapter has a categorized bibliography to direct further investigation of the topics discussed in the chapter.

I have attempted to provide all the information necessary for a reader who is unfamiliar with CAD to understand all the tools. Some readers, especially those who are already practicing CAD programmers, may find the presentation tedious and overly-detailed in places. I deemed it better to include the detail and allow the advanced reader to skim, rather than to leave out the detail and force the novice to dig.

Acknowledgements

Primary thanks go to the management and the employees of VLSI Technology, especially Al Stein, Henri Jarrat and Doug Fairbairn who kindly accepted this drain on my time.

Additional thanks are due to Jim Rowson, who assisted in the original outline five years ago and who provided daily encouragement and suggestions and who was often the first to read rough drafts. I wish to thank the VLSI Design Technology group who described algorithms, read initial drafts and corrected mistakes and misperceptions, especially Bill Walker, Ken Van Egmond, Tom Schaefer, Paul McLellan, Dick Lang and David Chapman. Thanks also to the internal reviewers, Tom Bulgerin, Suresh Dholakia, Nancy Gomes, Leslie Grate, Mike Grossman, Mark Hartoog, David Hsu, Chris Kingsley, Mike Kliment, Jim Lipman, Antonio Martínez, Scott Nance, Charles Ng, Shawn Purcell, Jim Rowson, Bill Salefski, Laura Smith, Russ Steinweg, and Bob Shur.

This book was formatted and printed entirely on VLSI Technology's documentation preparation system with figures and examples from the editors of VTItools. Many thanks to the people who developed the documentation system and kept it running, particularly Paul McLellan, Art Cabral, SE Tan, Debby Hungerford, Paul Gazdik and Flo Paroli.

An Introduction to CAD for VLSI

CHAPTER 1

INTEGRATED CIRCUIT DESIGN

We begin our investigation of design tools with a short discussion of integrated circuit design and the design process. We discuss the kinds of data that users manipulate and the tools they need during the design process.

Introduction to Integrated Circuit Design

Integrated circuit manufacturing consists of a series of steps, each of which adds or removes material from an area on a flat surface, typically of silicon. A designer describes a circuit as set of two-dimensional patterns on different *mask layers*, each of which represents one of the manufacturing steps. Other processing steps affect the entire wafer and the actual masks that are used in manufacturing may be combinations of the designer's mask layers, but we will take the design point of view and ignore these manufacturing details. We will consider only the layers that the designer uses to specify the chip.

In this section, we discuss the fundamentals of integrated circuit design. The purpose of this section is not to enable you to design circuits, but to provide you with an understanding of the basic tasks for which we develop tools. For more detail on design and manufacturing issues, see Mead and Conway (1980) or Mukherjee (1986). For a further discussion of design methodology, see Lattin (1979), vanCleemput (1979), Trimberger et al. (1981) or Niessen (1983).

Mask Layers

We describe the layers in a dual-well CMOS (Complementary Metal Oxide Semiconductor) manufacturing process. Different CMOS processes may use different layers, but the design issues are the same.

1. *Metal.* The primary wiring layer, typically aluminum.
2. *Polysilicon.* Poly-crystalline silicon, usually abbreviated *poly*, forms the gates of transistors. It is also a conductor and used for wiring.
3. *N+ Diffusion.* Silicon diffused with impurities to create a large excess of Negative charge carriers (electrons). It forms the source and drain of MOS N-channel transistors. It is a conductor and is used for wiring.
4. *P+ Diffusion.* Silicon diffused with impurities to create a large excess of Positive charge carriers (electron holes). It forms the source and drain of MOS P-channel transistors. It is a conductor and is used for wiring.
5. *N-Well.* A region of the silicon that has been doped to have a small excess of Negative charge carriers. It surrounds P-channel transistors.
6. *P-Well.* A region of the silicon that has been doped to have a small excess of Positive charge carriers. It surrounds N-channel transistors.
7. *Contact Cut.* A hole cut in the insulating oxide to allow metal to connect electrically to polysilicon and the diffusions.
8. *Overglass Cut.* A hole cut in the protective overglass layer to allow bonding wires to connect to metal bonding pads.

Designers use colors to distinguish the different layers, but there is no standard color scheme. Throughout this book, we will show pictures of circuit layouts in black and white using different patterns to represent different layers.

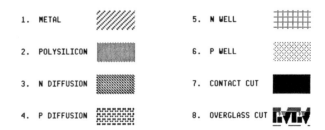

Wires

We can make wires in metal, polysilicon or the diffusions. Metal is preferred, because it has lower resistance and capacitance than the others, but polysilicon and diffusion may be used as well. We form a contact between metal and polysilicon with a contact cut. The structure including the metal, the contact cut and the polysilicon or diffusion is called a *contact*. An integrated circuit manufacturing process may have two or more metal layers separated by an insulating oxide. Designers have additional masks corresponding to the additional metal layers and connections between them, called *vias*.

Transistors

A polysilicon wire cuts a diffusion wire where they cross forming a MOS transistor. The polysilicon becomes the gate of the transistor and breaks the diffusion into two electrical nodes, the transistor source and drain. There are two kinds of transistors in CMOS. If we form a transistor on P-diffusion in an N-well, we get a *P-channel transistor*. When we form a transistor on N-diffusion in a P well, we get an *N-channel transistor*.

A MOS transistor acts like a switch. When there is a high voltage on the gate of an N-channel transistor, the switch is closed, making the connection between the source and drain. When there is a low voltage on the gate, the switch is open and source and drain are separated. A P-channel transistor works the same way except that a high voltage opens the switch and a low voltage closes it. In NMOS, there is a third kind of transistor, a depletion transistor. A *depletion* transistor is always closed, but has high resistance and is used as a resistor.

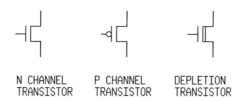

N CHANNEL   P CHANNEL   DEPLETION
TRANSISTOR   TRANSISTOR   TRANSISTOR

Electrical Details

An N-channel transistor turns on when there is a high voltage on its gate with respect to the substrate under the transistor. During design, a designer must connect the area under the N-transistors to VSS and the area under the P-transistors to VDD to provide the reference voltage for the transistors. These connections are called *substrate contacts* or *well ties* and not only allow transistors to operate correctly, but also prevent a catastrophic failure of CMOS circuits called *latchup*. Integrated circuit designers will certainly notice the absence of substrate contacts in all examples in this book. They have been intentionally omitted to enhance the clarity of the presentation of the tools.

Logic Gates

In MOS technologies, the high voltage power supply is called *VDD*, the low voltage is *VSS* or rarely, *GROUND*. We build logic gates in CMOS by making a *pullup structure* with transistor switches that connect the output to VDD when we want the output to go high and a *pulldown structure* that connects the output to VSS when we want it to go low. We make a CMOS inverter like the following figure.

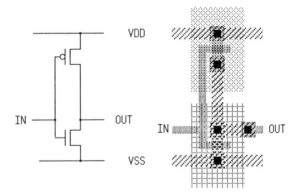

VDD is connected to the metal wire on top, VSS is connected to the wire on the bottom. The input signal enters in polysilicon on the left and forms the gate of two transistors, an N-transistor on the bottom and a P-transistor on the top. The bottom N-transistor turns *on* when IN is high, connecting OUT to VSS. The top P-transistor turns *on* when IN is low, connecting OUT to VDD.

Connections in series make AND-type connections, connections in parallel make OR-type connections. To connect the output to VSS when two signals are both high, we connect two N-transistors in series to VSS. To connect the output to VSS when either of two signals are high, we connect two N-transistors in parallel to the output.

We can make more complex gates by connecting transistors in series in the pullup structure and parallel in the pulldown structure (and vice versa). We make a NAND gate by making a series structure to VSS and a parallel structure to VDD. The output goes low if both A and B are high, and goes high if either A or B are low.

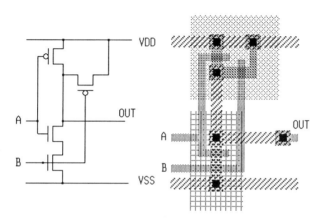

We use only N-transistors in the pulldown structure and only P-transistors in the pullup structure because N-transistors pass VSS well and P-transistors pass VDD well. The result is that we build inverting gates, NAND and NOR instead of AND and OR. To build an AND gate, we build a NAND gate and invert the output.

Electrical Characteristics

A layer of wire and the surface of the silicon substrate form a parallel-plate capacitor. All wires and transistors have capacitance to the substrate proportional to their area. The edges of layers also have a capacitance to the substrate proportional to the length of the perimeter. The amount of capacitance per unit area and per unit length of perimeter depend on processing details.

Every wire and transistor also has a resistance that is proportional to the ratio of its length to its width. Resistance is given in ohms per square, without regard to the size of the square. A wire one micron wide and one micron long has the same resistance as a wire one centimeter wide and one centimeter long.

The source of delay in integrated circuits is the time it takes to charge and discharge the capacitance on electrical nodes through the resistance of transistors and wires. We will estimate the delay as the product of the capacitance to be charged and the resistance through which it must be charged:

$t = RC$

Usually, the resistance of transistors dominates the resistance of wires, but because wires can be very long, the capacitance of wires is often larger than the capacitance of the transistors on those wires.

The Integrated Circuit Design Process

The design of one transistor or even a few transistors is simple, but the design of a system incorporating hundreds of thousands of transistors is not. In this section, we follow the flow of a design from inception through completed design, discussing the kinds of information that we track and the kinds of tools that designers use to handle that information.

Design Refinement

The design process starts with a rough *behavioral* description of the chip, in which the designer codes the algorithms he will use to perform the overall function needed by the part. He refines the behavioral description to a *structural* or *circuit* description, which includes logical connections between functional sub-blocks. Finally, the designer develops the *physical* description of the chip, which includes the mask layout.

Behavioral Design

The initial specification of a chip is a description of its behavior. The designer writes the function he wants in a *behavioral language*. A behavioral language is similar to a program and, in fact, the behavioral description language may be a programming language.

A designer verifies the behavioral description by simulating it. If the behavioral language is a programming language, the designer may simulate it by compiling and executing the program. However, a sequential program may be insufficient to express the parallelism inherent in an integrated circuit, so a designer may use a *behavioral simulator* with its own language instead of a programming language. A behavioral language that includes the ability to describe timing is often called a *hardware description language*.

During behavioral design, the designer decomposes the behavioral description into pieces to facilitate coding and understanding. The pieces do not necessarily represent partitioning of the function into different parts of a chip.

Circuit Design

At some point, the designer *partitions* the blocks in the behavioral description into units that represent pieces of the implementation. The connections between the blocks represent communication among those pieces. This is a structural description that represents *circuit* connections.

The simplest form of a structural description is *netlist*. A netlist is an interconnection of cells that represent the implementation of the design. The connections represent logical connections among the blocks, but not physical placement. The cells in the netlist may be as small as transistors or as large as complete processors.

A designer uses a *schematic editor* to enter a design as a netlist. A schematic drawing contains all the interconnection information in the netlist plus positions of all features for display and graphical *icons* to represent the blocks.

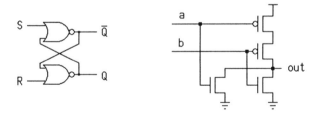

A common form of netlist represents the connection of *gates* representing boolean logic functions such as AND, OR and NOT. Thus, this phase of the design process is also known as *logic design*. However, the netlist may be more detailed, specifying a circuit in terms of transistors, resistors and capacitors.

A designer verifies the circuit with a simulator. The type of simulator depends on the kinds of cells in the netlist. If a designer developed a netlist of gates, he may use a *gate level simulator* to simulate operation of the circuit in terms of those boolean operations. Similarly, there are *transistor simulators*, also called *switch simulators* for a transistor-level netlist and *circuit simulators* for highly-accurate simulation of netlists of transistors, capacitors and resistors. A gate level simulator or transistor simulator that simulates accurate timing is called a *timing simulator*. A simulator that provides a combination of simulation modes is called a *mixed-mode simulator*.

At this point in the design, a designer usually checks the delays on critical paths through the circuit. He can run his netlist in a timing simulator to verify the timing or he can use a *timing analyzer* to get an estimate of the path delays without simulation.

Manual Physical Design

During physical design, the designer assigns physical structures to the logical functions in his circuit. Physical design starts with *floorplanning*, during which a designer works out the overall physical layout of the chip. During floorplanning, the designer estimates areas and aspect ratios and looks at critical timing paths. Typically, a few important cells are designed and simulated during floorplanning to get those estimates. After floorplanning, the designer knows the relative positions of major blocks in the system as well as their approximate sizes.

Finally, the designer builds and assembles the *layout* to implement the logic he has described. The layout, also called the *artwork* or *mask geometry*, consists of two-dimensional patterns on layers that represent the masking patterns for the integrated circuit manufacturing steps.

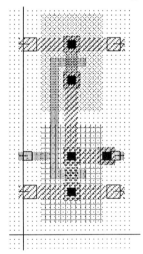

The most common physical design tool is a *layout editor*. A layout editor lets a user draw graphical shapes on different layers to define the mask layout. A layout editor does not check the correctness of those shapes, so a designer must run a battery of checking software to ensure that the layout he has made is correct.

A *symbolic layout editor*, or *sticks editor*, restricts the kinds of shapes a user can draw to a small set of symbols, such as transistors, contacts between wiring layers and wires on the wiring layers. Symbolic layout eliminates a class of layout errors but at a cost in layout density.

Procedural layout is another common form of physical design. In a procedural layout system, the designer expresses the design in a programming language with procedure calls to make graphical objects. The procedural style lets the designer use programming techniques to lay out the chip, including the use of conditionals, loops, variables and procedure calls.

Physical Design Analysis

Physical design of integrated circuits is detailed and error prone. Geometrical *design rules*, derived from the resolution and alignment tolerance of photolithography, dictate the minimum feature sizes, spacings between features and overlaps to ensure connection. Failure to meet any one of these rules anywhere on the chip may make the chip inoperable, unreliable or unmanufacturable.

Integrated circuit manufacturing is expensive, both in terms of money and time. The cost of manufacturing a faulty chip is much greater than the price of checking it. A designer employs a battery of checking tools to verify the correctness of the design.

A *plotter* is the most common design analysis program. The plotter draws the layout layers in different colors and patterns on a graphics display or on a piece of paper. The designer checks the plot by eye. A much more reliable tool for finding design rule errors is a design rule checker. A *design rule checker*, always abbreviated *DRC*, checks the sizes and spacings of the layout to ensure that they meet the design rule limits.

A *circuit extractor* builds a netlist of transistors from analysis of the layout. This is a translation from the physical description back to a structural description. After extraction, the designer can simulate the physical layout with a switch simulator or circuit simulator.

A designer can run other checking programs to verify the extracted netlist. A *network comparison* program compares two netlists for equivalence. This is used most commonly to check that the netlist generated during circuit design matches the extracted netlist from physical design. An *electrical rules checker* or *ERC* checks the netlist to ensure that the electrical properties of a circuit are not violated. These checks may include checks that all nodes could be driven both high and low, that power and ground are not shorted, that transistors do not drive ridiculously large loads and so on.

Automated Physical Design

In an effort to circumvent the large amount of checking necessary with manual physical design, *design automation* tools have been developed to automate the translation to the layout. These tools accept a behavioral or structural description of the chip and build the layout automatically. The layout is constructed with a set of predefined cells in a fixed floorplan. The resulting layout is usually not as efficient as a human designer would make, but it is finished quickly and it avoids most of the errors of manual physical design.

The most important tools for automated physical design are placement and routing tools. A *placer* maps the cells in the netlist to predefined physical gates and assigns those gates to positions on the chip. It chooses the positions by trying to optimize some criterion, such as the total length of the wires needed to connect the cells.

A *router* finishes the layout by generating the wires that connect the cells. There are several kinds of routers, the most common are a channel router and a maze router. A *channel router* routes all wires in a rectangular, or nearly-rectangular channel simultaneously. A *maze router* or *Lee-Moore router* routes wires one at a time on a grid. The grid can have barriers called blockages. The algorithm searches for a path around the blockages. When a wire is finished, the router adds the new wire's path to the list of blockages.

There are three major types of floorplans for automatic physical design using placement and routing: *standard cell, gate array* and *arbitrary block*. All three floorplans start with a netlist connecting a set of pre-defined cells. In a standard cell system, the cells are assembled into rows that are separated by variable-sized wiring areas. In a gate array, the cells are defined in terms of connections among pre-fabricated transistors. The placer makes its placement by customizing the bare transistors with a predefined overlay of connections that build logic gates. The transistors lie in rows and the system routes the connections between the rows. In an arbitrary block system, the cells can be any size and connections are made in the spaces between them. All three tools use a combination of a channel router and maze router to make the final connections.

A *PLA generator* is a program that generates a Programmable Logic Array (PLA) from a description of its function. The PLA is a regular structure for building arbitrary two-level logic. A *logic optimizer* or *PLA optimizer* reorganizes the logic in the PLA to reduce its area without changing its function.

A *silicon compiler* translates a behavioral description of a chip into a physical description. A silicon compiler assumes a particular regular structure, such as a microprocessor datapath, so it performs well on those structures, but is particularly ineffective for designs that do not fit its floorplan. A silicon compiler may use placement and routing or PLAs to implement the function.

Testing

Testing is usually the last issue addressed in integrated circuit design. Testing an integrated circuit refers to checking completed parts to ensure that they were manufactured correctly. The designer develops a test as a set of *test vectors*, which are the stimulus and expected response for a chip. Test vectors are usually generated by hand using a text editor and expressed in a *test language*. The test language lets a designer specify stimulus for the circuit, expected responses and limiting delays.

A *test generator* automatically derives tests for a circuit. Test generators usually require some restrictions on the design. A *fault simulator* grades tests by simulating the circuit assuming a fault and verifying that the test vectors propagate the differences due to that fault to the outputs. The percentage of faults detected gives an indication of the quality of the test.

Variances from This Model of the Design Process

This is a generalized description of the integrated circuit design process. The actual process used to design chips does not usually follow such an orderly progression, and large variances to this design flow are common. These variances result from necessity and convenience and do indeed produce working chips. To be successful, a tool must address a problem in the design process that designers use, so listen to the people who use your tools or, better yet, use the tools yourself to make chips.

- To make good decisions about which algorithms and structures to use, a designer must understand how those algorithms map to the physical description, so some trial and error is necessary.
- Design *iterations* result when a designer discovers problems with structure or behavior after those design phases are done. The design flow cycles back to higher levels for re-work. Design iterations may be so common and so large that the separation of design tasks may be impossible.
- Work may progress on the behavior, structure and physical design simultaneously. The design of different parts of a chip may proceed at different speeds, and time pressure to complete the design usually requires that many phases of the design proceed in parallel.
- A typical integrated circuit design project has many designers working on different parts or different phases of the design simultaneously. The different parts of the design may be at different phases of the design process or at the same phase. These designers face problems of verification and maintenance of interfaces and coordinated multiple access to cells.

Managing the Complexity of a Design

Cells and Instances

Just as it is too difficult to design a single procedure with a hundred thousand lines, it is too difficult to design a chip with a hundred thousand transistors. Instead, a designer divides the design along functional boundaries, as we have mentioned, into smaller pieces called *cells*. He subsequently divides those cells into even smaller cells until the problem is manageable. Each cell is composed of primitive objects and references to other cells called *instances*. This structure of instance references and cells gives rise to a hierarchy of cells called the *design hierarchy*.

For example, we can have a microprocessor that consists of a microsequencer, a data processing section and microcode memory. The data processing section can be divided into an ALU, registers, shifter and multiplexers.

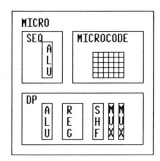 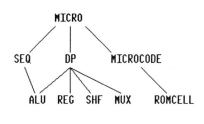

Expressing each cell in terms of instances can vastly reduce the space needed to store the design. Even though we use a cell many times in a design, we only need store it once. We can edit a small portion of the chip without being forced to load the whole chip into memory, nor need we concern ourselves with accidentally modifying part of the chip we do not wish to change. A correction to a cell is reflected in all its instances, so we can be confident that improvements propagate to all parts of the design. Just as procedures and modular programming make software easier to understand, hierarchy makes a chip easier to understand by hiding large amounts of unnecessary detail. Hierarchy does not solve all problems, however. Some designs do not have a straightforward hierarchical decomposition, and some parts of the design process, such as simulation, may require that we flatten the hierarchy.

All our software deals with cells. Whenever a user enters a design, it is contained in a cell. The complete chip is a single cell with instances representing pieces.

Types of Cells

A cell has a type that indicates the kind of data in the cell. A layout cell has information regarding the placement of rectangles and polygons for the physical artwork of the chip. A circuit netlist cell has information about logical connections among transistors in the design. Each cell includes both primitive structures peculiar to its type (polygons in a physical description, transistors and nodes in a circuit description, and so on) and instances to reference other cells.

Hierarchical references are different with different data types. An instance of a layout cell must not only name the cell but also give the orientation of that cell in the larger cell. An instance of a circuit cell has no orientation requirement, but must specify which electrical connections must be made between the cell and the environment of its instance.

Regularity

The second major method used by designers to reduce the complexity of design is regularity. A design that is composed of an array of identical structures is easier to understand than one made up of scattered, unique structures. Further, regular structures with structured interconnect reduce the space needed for wiring, so the chips are smaller and run faster.

Methodology

Tools that are built around hierarchical design force designers to use a hierarchical design methodology. This restriction is not inherently evil, but you should be aware that decisions made for the purposes of tool efficiency or design clarity affect the allowable design styles. In this case we buy a method to reduce the complexity of the design. However, this advantage may not be useful to some designers who are not building complex designs, so the restriction would simply be a burden.

Another design restriction that has been proposed for simplifying design tools is restricting layouts to *orthogonal* angles, only horizontal and vertical lines. The design-tool advantages from this simplification are large: all distance and comparison calculations are very simple. Design rule checking is easy and fast, and we can use symbolic layout tools to ensure automatic design rule correctness. However, eliminating non-orthogonal angles in layouts makes the layout larger. Larger designs are more expensive to manufacture. The tool development cost occurs once, the manufacturing cost occurs every time the chip is made. Therefore, the savings due to arbitrary-angled layout has dominated the savings for simpler tools and the restriction has been deemed unacceptable by the industry.

We will see other design restrictions as we go. Since computer aided design is an application area, it is the user of the tool that must decide whether or not the savings is warranted, not the tool builder. The key thought to keep in mind when you consider adding a design restriction is what the restriction buys for the designer and at what cost. Designers do not value highly the internal simplicity of tools.

Exercises

1. Estimate the amount of storage needed for a one-hundred-thousand transistor chip. Break down your estimate for behavioral, circuit and physical descriptions. Make one estimate assuming no hierarchy and one with a hierarchy branching factor of ten (each cell includes ten instances).

2. Describe techniques currently used managing the complexity of software design. Do these techniques also apply to integrated circuit design?

3. What tools are there for software design that are analogous to those used for integrated circuit design? Why are there some integrated circuit design tools that have no analogous software design tools?

References

B. Lattin, "VLSI Design Methodology: The Problem of the 80's for Microprocessor Design", *Proceedings of the Caltech Conference on VLSI*, 1979.

C. Mead and L. Conway, *Introduction to VLSI Systems"*, Addison-Wesley, 1980.

A. Mukherjee, *Introduction of nMOS and CMOS VLSI Systems Design*, Prentice Hall, 1986.

C. Niessen "Hierarchical Design Methodologies and Tools for VLSI Chips", *Proceedings of the IEEE*, vol. 71, January 1983.

S.M. Trimberger, J. Rowson, J.P. Gray and C.R. Lang, "A Structured Design Methodology and Associated Software Tools", *IEEE Transactions on Circuits and Systems*, vol. CAS-28, no. 7, July, 1981.

W.M. vanCleemput, "Hierarchical Design for VLSI: Problems and Advantages", *Proceedings of the Caltech Conference on VLSI*, 1979.

CHAPTER 2

PARSING

Reading Files

There are two parts to reading a data file: processing the characters and building the results into a data structure. The former is commonly called syntax analysis and the latter semantic analysis. Since the data structures we wish to build may differ with every tool, we separate the two jobs. A *parser* is the tool that processes characters in an input stream and emits procedure calls to a module that implements the semantics for the current tool.

This chapter includes information about parsers with emphasis on a particular type called a recursive descent parser. The information in this chapter can also be found more formally and more generally in books about parsing, such as Aho and Ullman (1972 and 1977), but we will limit our coverage to only that part of the problem necessary to generate a parser that is applicable to our needs.

This chapter is tutorial in nature, describing an implementation. In this chapter, we develop a parser that is used in tools in later chapters. The parser is useful not only for reading data files, but also for processing user input.

Describing a Data Format

A data format is a language. A language definition consists of a *syntax* and a *semantics*. The semantics is what you say, the syntax is how you say it.

Syntax Description

We express the syntax of a language in a notation called *Backus-Naur Form*, or *BNF*. The following BNF describes a data format for layout information. Each equality is called a *production rule* or just *rule*. Names on the left side of the equals sign are called *nonterminals*, they are defined in terms of other nonterminals and *terminals*, the strings in quotes. Terminals appear in the input exactly as they are written. Curly braces indicate zero or more repetitions, square brackets surround optional text and vertical bars indicate choices. Parentheses may be used for grouping.

layoutFile	= version { layoutCell | commentStatement } endFileStatement
version	= "V" integer ";"
layoutCell	= cellHeader cellBody
cellHeader	= cellStatement boundingBox connectors
cellStatement	= "CELL" name name "layout" ";"
boundingBox	= "G" point point ";"
connectors	= { connectorStatement | commentStatement }
connectorStatement	= "C" name integer real point integer ";"
cellBody	= { bodyStatement } endStatement
bodyStatement	= layerStatement | boxStatement | polygonStatement | instanceStatement | commentStatement
layerStatement	= "L" integer ";"
boxStatement	= "B" point point ";"
polygonStatement	= "P" { point } ";"
instanceStatement	= "I" name point integer name ";"
commentStatement	= "#" { any character except ";" } ";"
endStatement	= "E" ";"
point	= real real
endFileStatement	= "ENDFILE"

A `layoutFile` consists of a `version` followed by any number of either a `layoutCell` or a `commentStatement`. A `version` consists of the letter "V" followed by an integer and a semicolon. We could go on with the BNF, defining `name`, `integer` and `real` as sequences of characters and digits. Instead we say the BNF deals with tokens separated by white space. White space consists of blanks, tabs and newline characters. A *token* is a name, an integer, a real number or a special character. The part of a parser that generates tokens is called a *lexical analyzer* or *token scanner*.

Semantic Description

There is no such generally-used precise method for describing the semantics of the data form. Instead, most of the time we rely on a natural language to accurately describe each statement.

Numbers in the layout file are in units of *lambda* (λ) (Mead and Conway 1980). λ is a scale factor that is a measure of the resolution of the manufacturing process. As the manufacturing technology improves, λ becomes smaller. We can use the same layout file with the improved technology without changes. We need not concern ourselves in the tools with the size of λ. All our tools will use it as the fundamental unit of length and we will do all our calculations internally in units of λ and scale our layout files to physical units at the end of the design process.

Some nonterminals form complete objects. We describe the semantics of those nonterminals by describing the objects they represent. We introduce each statement with a BNF-like description of its statement fields and give an example if its use.

version   = "V" versionNumber ";"

`V 1;`

The version statement gives the version number of the layout file format used in this file. We can have our software check the version number before reading so we can maintain compatibility of tools yet allow the format to develop.

cellStatement   = "CELL" cellName technologyName "layout" ";"

`CELL register CMOS layout;`

The cell statement starts a cell definition, providing the cell name and the name of the technology in which the cell is defined. So far, the only kind of cell we have seen is layout, so we label this cell as a layout cell. In Chapter 8, we will see cells of netlist type. By including a cell type, we will be able to combine cells of different types in the same file and have our software ignore cells of the wrong type.

boundingBox   = "G" leftBottom rightTop ";"

G 0 0 45 50;

The *bounding box* of a cell is the smallest rectangle aligned on the x-y axes that includes all layout in the cell.

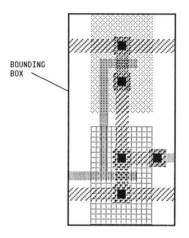

connectorStatement = "C" nodeName IDnumber width location layer ";"

C vdd 1 12 45 20 1;

The connector statement identifies a location where a wire should connect to the cell from the outside. The node name is a name for identifying the connector. The ID number is a unique identifier for the connector, since there may be more than one connector on the same electrical node. The **width** is the width of the connector along the edge of the cell and the **location** is the location of the center of the connector on the cell. The layer number gives the layer of this connector.

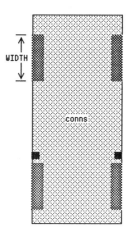

layerStatement   = "L" layer ";"

L 1;

The layer statement sets the layer for later box statements and polygon statements. Requiring a number for a layer may seem obtuse. However, since this data form will be read and written by machine, we need not concern ourselves with human readability. We use numbers because they are simple for machine processing. We translate numbers to names if people must read the data. We will use the following number assignments for our CMOS examples:

1. Metal
2. Polysilicon
3. N diffusion
4. P diffusion

5. N-well
6. P-well
7. Contact cut
8. Overglass cut

boxStatement   = "B" leftBottom rightTop ";"

B 1 1 15 3;

This box statement describes a box from coordinate 1,1 to 15,3. The box is two units high and 14 units wide. The box statement must be preceded by a layer statement to set the layer of the box.

polygonStatement   = "P" { point } ";"

P 1 1 1 5 3 4;

This polygon statement describes a triangle. It must be preceded by a layer statement to set the layer of the polygon.

instanceStatement   = "I" instName origin orientation cellName ";"

I i3 20 30 0 inv;

The instance statement indicates a placement of an instance of the cell named **cellName** with the origin of coordinates of that cell placed at the location specified. The cell must have already been defined in the file. The cell is then mirrored and rotated as specified by **orientation**. The instance has its own name, given as **instName**.

The orientation number is an integer in the range 0-7, representing the eight combinations of mirroring and rotation:
    0. No mirroring, no rotation.
    1. No mirroring, rotate 90 degrees counterclockwise.
    2. No mirroring, rotate 180 degrees.
    3. No mirroring, rotate 270 degrees counterclockwise.
    4. Mirror across Y axis, no rotation.
    5. Mirror across Y axis then rotate 90 degrees counterclockwise.
    6. Mirror across Y axis then rotate 180 degrees.
    7. Mirror across Y axis then rotate 270 degrees counterclockwise.

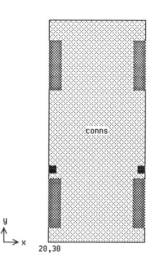

commentStatement = "#" { any character except ";" } ";"

this is the start of the pulldown

The comment statement lets users or programs insert comments without changing the layout. Comment statements are ignored when the cell is read.

endStatement = "E" ";"

E;

The endStatement marks the end of a cell.

endFileStatement = "ENDFILE"

ENDFILE;

The end file statement marks the end of the layout file.

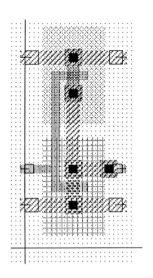 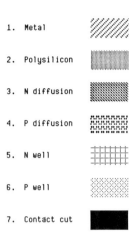

An Example

The preceeding figure shows a CMOS inverter and layer numbers for the layers. We can describe the inverter with this layout file:

```
V 1;
CELL inv cmos layout;
G -1 2.5 23 49.5;
C vddL 1 3 0 42.5 1;
C vddR 2 3 22 42.5 1;
C vssL 3 3 0 9.5 1;
C vssR 4 3 22 9.5 1;
C in 5 2 0 13.5 3;
C out 6 2 22 17.5 3;
L 1;
   B -1 41 23 44;
   B -1 8 23 11;
L 4;
   B 9 40.5 13 44.5;
   B 9 32.5 13 36.5;
L 1;
   B 9 44 13 44.5;
   B 9 40.5 13 41;
L 6;
   B 10 41.5 12 43.5;
L 5;
   B 9 19 13 19.5;
   B 9 15.5 13 16;
L 6;
   B 10 16.5 12 18.5;
   B 10 33.5 12 35.5;
L 3;
   B -1 12.5 14.5 14.5;
   B 6 37.5 14.5 39.5;
L 4;
   B 9.5 36.5 12.5 40.5;
L 3;
   B 6 14.5 8 37.5;
L 5;
   B 9.5 11.5 12.5 15.5;
L 1;
   B 17 19 21 19.5;
   B 9 16 21 19;
   B 17 15.5 21 16;
L 6;
   B 18 16.5 20 18.5;
L 3;
```

```
    B 9 15.5 13 19.5;              B 17 18.5 21 19.5;
    B 9 7.5 13 11.5;               B 17 16.5 23 18.5;
L 1;                               B 17 15.5 21 16.5;
    B 9 11 13 11.5;            L 8;
    B 9 7.5 13 8;                  B 4 2.5 18 24.5;
L 6;                           L 9;
    B 10 8.5 12 10.5;              B 4 27.5 18 49.5;
L 1;                           E;
    B 9 32.5 13 36.5;
    B 9.5 19.5 12.5 32.5;      ENDFILE
```

Parser Overview

We will implement a *recursive descent parser*. A recursive descent parser has one procedure for each nonterminal in the BNF. When the BNF indicates a terminal, we check that the terminal is present and give an error if it is not. There may be more than one legal terminal because the decisions at iterations and choices in the BNF are signalled by terminals. When the BNF indicates a choice or the possible conclusion of an iteration, the parser checks one token ahead of its current position to make the decision. When the BNF indicates a nonterminal, we call a separate procedure to parse that nonterminal. When we have completed processing a statement, we call the semantics procedure to build the data structure for that statement.

This technique handles grammars of a type known as LR1, meaning they can be parsed reading from left to right looking ahead at most one token. These grammars are easy to read and are sufficient to express the kinds of information we have. The layout file format and many others we encounter, do not require the full power of a recursive descent parser, since their rules are not recursive. We could write a non-recursive state-driven parser to read it. We will discuss a recursive descent parser anyway because it is easy to implement and because it is can parse programming languages with recursive rules, a feature that will be useful when implementing more powerful languages, such as procedural layout, which we will see in Chapter 6.

We have a straightforward method of translation from a formal description of a syntax to a program to read that syntax. This task may seem mechanical, and it is so. In fact, automated parser generators have been written to translate BNF to a parser with no human intervention.

Implementation of the Parser

When reading a file, three modules interact: the client, the semantics and the parser. The semantics module has procedures that build data structures for the client. The client may also be the semantics module, but we will separate them so we can use the same semantics module in other applications. The client starts the parser to begin reading the file.

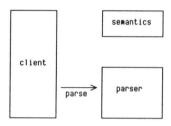

The parser calls the semantics procedures in the semantics module. Those procedures build the data structures for the client.

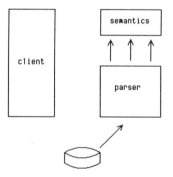

When the parse is finished, the parser returns to the client, which retrieves the data structure from the semantics module.

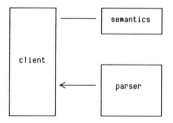

Parser Module Interface

The parser module has four global variables: the file we are reading, a token for the one-token lookahead, a pointer to the semantics module and an error flag. The parser has one externally-accessible procedure: parse, that takes the name of the file to parse and a pointer to the semantics module and returns *false* if errors were found during the parse.

```
module layoutParser(

  boolean procedure parse(
    string fileName;
    pointer(layoutSemantics) sem;
  );
);
```

Semantics Module Interface

The interface of the semantics module consists of one procedure for each of the complete statements in the language. Each procedure in the semantics module returns a boolean error condition to stop parsing. A single error should not stop a programming language compiler, which must operate as if its input was generated by a person and therefore full of errors. In our case, we are reading a machine-written data file that we assume to be correct. Any problem in the data file syntax is disastrous since it indicates an error in the software that wrote the file.

The semantics module has an interface that allows the client to retrieve the data structure that it builds during parsing. The interface may be another procedure that returns a pointer or it may be a field of the semantics module that is accessible from the client. In our case, the data we want is a list of cells. We use the term "list" here and throughout the book to represent a collection without specifying the implementation. For convenience you may consider them to be linked lists or arrays.

```
module layoutSemantics(
  pointer(list) cells;

  boolean procedure version(integer versionNumber);

  boolean procedure cellStatement(
    string cellName, technologyName);
```

```
  boolean procedure boundingBox(
    real left,bottom,right,top);

  boolean procedure connectorStatement(
    string nodeName;
    integer connectorIDnumber;
    real width,x,y;
    integer layer;
  );

  boolean procedure layerStatement(integer layer);

  boolean procedure boxStatement(
    real left,bottom,right,top);

  boolean procedure polygonStatement(
    real array (1 to *) xarray,yarray);

  boolean procedure instanceStatement(
    string instanceName;
    real x,y;
    integer orientation;
    string cellName;
  );

  boolean procedure endStatement;

  boolean procedure endFileStatement;
);
```

The Semantics As a Procedure

Many programming languages don't have a module structure or they don't allow modules to be passed as parameters as we pass the semantics module to the parser. We can achieve the same effect by passing individual semantics procedures as parameters to **parse**.

```
boolean procedure parse(
  string filename;
  boolean procedure version(integer versionNumber);
  boolean procedure cellStatement( ... );
  boolean procedure boundingBox( ... );

    ...
```

```
  boolean procedure endStatement;
  boolean procedure endFileStatement;
);
```

If our programming language does not allow either passing a module as a parameter or passing procedures as parameters, we can implement the semantics by hard-coding the semantic procedure names in the parser. However, this limits our ability to re-direct the calls to the semantics, a powerful technique we will discuss later in this chapter. If your language has these kinds of restrictions, you should consider using a different programming language.

Internal Parser Procedures

Mechanics of the Parser

For simplicity, we name the procedures in the parser the same as the nonterminal of the rule they parse. We have an additional procedure, **parse**, which we call with the name of the file to read and a pointer to the semantics module. **parse** sets the global **semantics** pointer, opens the file, reads the first token into the global **tok** and calls a procedure for the first nonterminal, layoutFile. Subsequent procedures follow the BNF as we described earlier. tok always contains the next token to be examined.

```
boolean procedure parse(
  string fn;
  pointer(layoutSemantics) sem;
);
begin
  errorFlag := false;
  semantics := sem;
  filePointer := open(fn,input);
  getTok;
  layoutFile;
  if filePointer<>nullpointer then close(filePointer);
  return(errorFlag);
end;
```

```
procedure commentStatement;
begin
  while tok<>";" do getTok;
  getTok;
end;

procedure cellHeader;
begin
  cellStatement;
  boundingBox;
  connectors;
end;

procedure layoutCell;
begin
  cellHeader;
  cellBody;
end;

procedure layoutFile;
begin
  if tok<>"V" then begin
    error("No version statement");
    return;
  end;
  getTok;
  version;
  while not errorFlag do begin
    if tok = "ENDFILE" then begin
      errorFlag := not semantics.endFile;
      return;
    end else if eof(filePointer) then begin
      return;
    end else if tok = "#" then begin
      gettok;
      commentStatement;
    end else layoutCell;
  end;
end;
```

When a procedure has semantics associated with it, like version does, it the appropriate procedure in the semantics module:

```
procedure version;
if not errorFlag then begin
  if not isNumber(tok) then begin
    error("Version Number is not a number");
    return;
  end;
  semantics.version(integerOf(tok));
  getTok;
  if tok<>";" then begin
    error("No semicolon following version statement");
    return;
  end;
  getTok;    # one token lookahead
end;
```

The Token Scanner

The token scanner, or lexical analyzer, builds the tokens and is embodied in the procedure getTok. We can write a token scanner as more BNF, describing token scanning on a character by character basis. However, many programming languages have built-in string processing procedures to assist the scanning task. Some languages include a scanner of the type we need. However, it may be necessary or convenient to implement a token scanner of our own.

In many ways, getTok looks like a small version of the parser we have just implemented: there is a global token, in this case a single character, that implements the one-character lookahead. The procedure checks that token and makes decisions on it. The tokens listed in the syntax description fall into three categories: words, numbers and special characters, so our token scanner has a three way conditional. The token scanner is useful for a number of applications: reading files, reading typed input from users and interpreting names or other strings internally.

```
procedure getTok;
begin
  skip white space
  if char is a letter then begin
    collect letters, digits and underscore into the token
  end else if char is a digit, "-" or "." then begin
    collect digits, decimal point and more digits
  end else begin
    token := char;
    char := readChar(filePointer);
  end;
end;
```

Building a Data Structure

The layout parser translates the declarations of graphical items into procedure calls. The layout semantics, which we discuss now, translates those procedure calls into data structures. First we introduce some simple data structures, then we describe a layout semantics that builds them.

As a general rule, we separate the data representing the state of a tool from the data of the cell on which the tool operates. This is not merely a programming convention, this separation of data and tool allows us to maintain the cell as a separate entity in our system. Thus, we can write new tools in the same program to work on the same data and have both tools modify the same cell, the same data structure in memory. For example, we could have a layout editor and a hardcopy plotter, written independently. We edit the cell in the layout editor and plot it with the plotter. Use of either tool does not affect the other tool. We can switch from one tool to the other without exiting the system or even copying the data. Further benefits of the separation between tool and data include the ability to enhance the cell's data without modifying the tool, and to manage the cell data in our computer's memory intelligently.

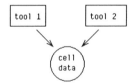

Data Structures

We use data structures that follow the layout file description. Each type of object has fields for its data. The semantics module connects data structures into a coherent whole.

Cells

Every cell has a name, a bounding box and contents. We make a list of all the graphical items in the cell, including all connectors, boxes, polygons and instances.

```
class cellClass(
   string name;
   pointer(boundingBox) bb;
   pointer(list) data;
);
```

Graphical Items

All graphical items include a bounding box. The bounding boxes are not part of the BNF specification for graphical items, we compute them when we read the cell and store them in the data structure. We use them to speed searches through the data.

```
class itemClass(
   integer type;
   pointer(boundingBox) bb;
);
```

The **itemClass** is a superclass of all graphical items. When we make a subclass of **itemClass**, the subclass inherits the **type** and **bb** fields. We define the superclass so we can have pointers to a data structure that tells us what kind of object we have. This allows us to put all our **items** in the same list. The type field differentiates boxes from polygons and so on. We can achieve a similar effect with variable records in Pascal.

A box item has a layer in addition to the bounding box. A polygon item has a layer and its **x** and **y** arrays. A connector includes a layer, a unique identifier number, a width and a name.

An instance has a name, used for identification purposes, the x,y position of its origin in the current cell, an integer orientation number and a pointer to its defining cell. The orientation and x,y position completely determine the mapping from the cell's coordinate space to the instance's.

```
define
  boxType       = 1,
  polygonType   = 2,
  connectorType = 3,
  instanceType  = 4;

class (itemClass) boxItem(
  integer layer;
);

class (itemClass) polygonItem(
  integer layer;
  real array (1 to *) x,y;
);

class (itemClass) connectorItem(
  integer layer,ID;
  real width;
  string name;
);

class (itemClass) instanceItem(
  string name;
  real x,y;
  integer orientation;
  pointer(cellClass) cell;
);
```

The Semantics Module

From the layout file, we build a data structure that consists of a list of cells. When the parsing is done, the client program retrieves the list from the semantics. Pictorially, the resulting data structure looks like this:

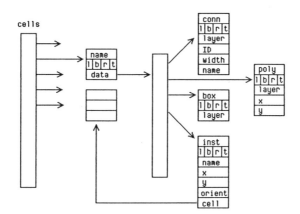

Finally, while building the data structure, we concern ourselves with error conditions. Just as the parser checked for syntax errors, the semantics module checks for semantic errors, such as a cell name or connector ID being re-used, a bad orientation number and so on. It is true that we do not expect such errors because users will not edit these files by hand, but there are three reasons for inserting the checks:
1. error checks are not unduly time consuming
2. they are invaluable while debugging
3. users modify files no matter what you tell them

Reading the File Header

The header on a layout file is just the version number. The semantics module checks the version number and if it is a version that it does not understand, it gives an error and returns false.

Reading a Cell Declaration

We build the cell data structure from the cell, bounding box and connector statements. Each cellStatement starts a new cell, so we allocate a new cellClass for it and make it the current cell.

```
boolean procedure cellStatement(
   string cellName, technologyName);
begin
   if currentTechnology<>technologyName and
      currentTechnology<>"" then begin
      error("Incompatible technology");
      return(false);
```

```
  end;
  if currentCell<>nullpointer then begin
    error("CELL statement inside a CELL");
  end;

  if lookupCell(cellName)<>nullpointer then begin
    error("Duplicate Cell name: "&cellName);
  end;
  currentTechnology := technologyName;
  currentCell := new(cellClass);
  currentCell.name := cellName;
  currentLayer := noLayer;
  return(true);
end;
```

The boundingBox semantic procedure fills in the boundingBox field of currentCell. Procedure endStatement terminates the additions to the current cell and adds the current cell to the cells list.

```
boolean procedure endStatement;
begin
  if currentCell=nullpointer then begin
    error("END cell with no CELL");
  end;
  putInList(cells,currentCell); # add the cell to the list
  currentCell := nullpointer;
  return(true);
end;
```

Reading Cell Contents

The cell's data list contains boxes, polygons, connectors and instances. When the semantics module receives a call for a primitive data object such as a box, polygon or connector, it creates the corresponding record, filling it with the data from its parameters, and adds it to the list of data in the current cell. The layer for boxes and polygons comes from the current layer set by the layerStatement procedure.

The following is code for the boxStatement. The box dimensions are stored in the bounding box field, bb. The polygon and connector procedures are similar to the box procedure. The bounding box for polygons and connectors must be calculated when the record is filled. In addition, we recompute the bounding box of the cell as we read the cell to verify that it is correct.

```
boolean procedure boxStatement(real left,bottom,right,top);
begin
   pointer(boxClass) b;
   if currentLayer=noLayer then error("No Layer for BOX");
   b := new(boxClass);
   b.type := boxType;
   b.layer := currentLayer;
   b.bb := new(boundingBox);
   b.bb.left := left;
   b.bb.bottom := bottom;
   b.bb.right := right;
   b.bb.top := top;
   putInList(currentCell.data,b);
   return(true);
end;
```

When we find an instance, we look up the cell we are instantiating and compute the instance bounding box by transforming the cell's bounding box. We will discuss transformations in Chapter 3. For now, we will assume that we have a procedure that transforms a box given the new orientation and displacement.

```
boolean procedure instanceStatement(
   string instanceName;
   real x,y;
   integer orientation;
   string cellName;
);
begin
   pointer(instanceClass) inst;
   inst := new(instanceClass);
   inst.type := instanceType;
   inst.name := instanceName;
   inst.x := x;
   inst.y := y;
   inst.orientation := orientation;
   inst.cell := lookupCell(cellName);
   if inst.cell=nullpointer then begin
      error("CELL "& cellname & "not found.");
   end;
   inst.bb := transformBox(inst.cell.bb,orientation,x,y);
   putInList(currentCell.data,inst);
end;
```

Memory Usage Issues

The layout of a large chip can be very large and may not fit within the limits of virtual memory of our computer. If we build the data structure in memory as we have described, we may overflow memory and the tool will crash. A very simplistic "solution" to this problem is to document the fact that the tools will not function with large chips. *This solution is unacceptable anywhere but in a learning environment.*

The most obvious resolution to this problem requires that we run our own virtual memory for our cells. We write whole cells out to disk when our memory is nearly full and bring them back into memory when we need them. If we do this, we cannot merely follow the `cell` pointer from an instance and expect that the data for the cell be there. We first check that the cell is in memory, and retrieve it if it was not. If there is insufficient memory for the cell, we remove other cells from memory. Readers who are familiar with operating systems principles will notice the similarity between these operations and a demand paged virtual memory. Our "pages" are cells.

We call the tool that manages cells in memory a *cell manager*. The cell manager can have many more functions than memory management. Since the cell manager decides which cells are in memory, a tool that edits a cell must first request it from the cell manager, which retrieves it if necessary. The cell manager can distribute duplicate pointers to a cell, so more than one tool can edit the same cell at the same time. The cell manager can also manage versions and control restricted-access libraries of cells.

Writing Layout Files

We can make a general layout file writer using the layout semantics interface. To write a file, a program mimics a parser, calling the semantics procedures in the writer. The advantage of this organization is that we can replace the layout writer semantics with some other layout semantics to redirect the output to another program without actually creating a file and without changing our code.

For example, if we have a layout editor and a design rule checker. The layout editor has a layout cell writer and the design rule checker has a layout cell reader. Normally, the layout editor writes files to disk and the checker reads from disk. We can substitute the checker's reading semantics for the editor's writing semantics and instruct the layout editor to write the cell. Instead of calling the layout file writer semantics, the editor calls the DRC reader semantics, building the DRC's data structures.

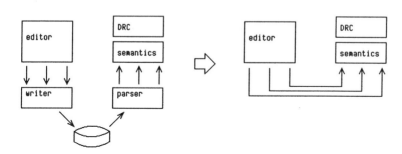

Exercises

Programming Problems

1. Finish the parser for layout files.

2. Change the parser to accept commentStatements between tokens anywhere in the input text. Hint: you need only change one procedure.

3. Write a translator to and from Caltech Intermediate Form (CIF). CIF is described in Mead and Conway (1980).

4. Write a semantics that types back the cell in a human-readable format, including mnemonics for layers and orientations.

5. Write a semantics that writes layout files.

6. Write a semantics that checks the bounding box on the cells in the file.

7. We ignored the type field on a cell, saying it will always be layout. Should the change to skip cells of the wrong type be made in the parser or in the semantics? Make the appropriate changes your parser.

8. Write a calculator program that lets you type expressions with addition, subtraction, multiplication and division. Multiplication

and division should happen before addition and subtraction. Your calculator should implement operator precedence and parentheses.

Questions

1. In the semantics section, we say that there must be a `layerStatement` before the first `boxStatement` or `polygonStatement`. Rewrite the BNF to reflect this restriction. What changes must you make in the code? In which procedure would this error be caught originally? Where would it be caught with your change?

2. According to the BNF, between which two statements in the layout file is it illegal to have a comment statement? Rewrite a portion of the BNF to allow a comment there.

3. Modify the BNF so bad orientation numbers are caught in syntax analysis. Is it reasonable to phrase the BNF to catch as many errors as possible, or is it better to let the semantics catch some?

4. Write the BNF for an expression evaluator. The expression evaluator should be able to handle four-function arithmetic with operator precedence and parentheses.

5. How would you change the parser to accept end-of-line rather than semicolon as the end of a command?

6. What are the advantages and disadvantages of implementing the `cells` list in our data structure as an array? A linked list? What are the advantages and disadvantages of implementing the **data** list in `cellClass` as an array or a linked list?

7. What error checks are required in the semantics module for boxes and polygons?

8. What modification to semantics procedures must you make if you are to implement a cell manager to control memory usage? What modifications would you like to make to the layout file format?

9. What are the advantages and disadvantages of keeping a layout file as binary instead of text?

References

Parsing

A.V. Aho and J.D. Ullman, *The Theory of Parsing, Translation, and Compiling Volume I: Parsing*, Prentice-Hall, 1972.

A.V. Aho and J.D. Ullman, *Principles of Compiler Design*, Addison-Wesley, 1977.

General

C. Mead and L. Conway, *Introduction to VLSI Systems*, Addison-Wesley, 1980.

CHAPTER 3

GRAPHICS

This chapter is a review of the subset of interactive graphics programming techniques we need to build graphical CAD software. A full treatment of interactive graphics is beyond the scope of this book, but you can find it in Newman and Sproull (1979) and Foley and Van Dam (1982). Some familiarity with interactive graphics techniques and terminology is assumed. If you are already familiar with interactive graphics or if you already have a graphics package, you may safely skip this chapter. However, it is recommended that you at least scan the chapter, since it provides a summary of the graphic primitives we use later. The sections on color and dynamic displays are of broader interest, since they are not usually included in graphics courses, yet they are important in our application.

Output Devices

We call graphical softcopy devices *displays* and graphical hardcopy devices *plotters*. We distinguish between two kinds of output devices: *raster* devices and *vector* devices.

Raster Devices

A raster type device displays an image as a grid of picture elements, *pixels*. In a black and white display, each pixel consists of one bit. In a color display, each pixel consists of more than one bit. A common color display for integrated circuit design consists of 1024 pixels square and has eight bits per pixel. In display devices, the color number in the pixel is commonly used as an index into an array called a *color map* to choose the actual color on the display for that pixel. The entries in the color map have many more bits than the pixel in the display memory, perhaps 24 bits. These bits are divided into three eight-bit fields that define the amount of red, green and blue in the final pixel value. Thus, the display can have as many as 2^{24} different colors, of which only 2^8 can be seen at any time. We choose which colors will be seen by loading appropriate values into the color map.

Raster plotters are typically higher resolution than raster displays, so they require more data. Further, they require that the data be ordered so the plotter does not have to rewind the paper. This ordering and the large amounts of data require internal buffering for the plotter.

Vector Devices

A vector device draws figures line by line. Vector plotters are common, but vector displays have been largely supplanted by raster devices. In a vector display, each line is re-displayed each time the display is updated -- usually thirty times per second. Therefore, changes to the data can appear within $1/30^{th}$ of a second, but the complexity of the data on a vector display is limited to the number of lines the display can draw in $1/30^{th}$ of a second.

A vector plotter draws each line separately, so the length of time to plot depends on the number of lines and on the distance the plotter pen must move between lines. Sorting the data can significantly shorten plotting time.

We prefer to display rectangles and polygons as filled areas. On raster displays, we set the colors of all pixels within the area we wish to show. On a vector display, we can fill them with hash marks, but this takes a long time, so for performance reasons, we usually outline graphical features on vector devices rather than fill them. Most raster devices and some vector devices provide hardware assistance for filling areas. Since we are usually concerned with interactive techniques, implementing display features in software that are not present in the hardware results in unacceptable run time performance.

There are a few displays with both raster and vector features. Such devices are very attractive, because we can display the bulk of the data with the raster facility and use the vector facilities for the few fast-moving parts of the display.

Graphical Output Primitives

The simplest form of graphical output consists of writing directly to a display device. Since every graphic device has its own control commands, We encapsulate that writing in a separate display *driver* module. We then define a universal graphics interface so we can draw graphics without having to deal with the command codes that are peculiar to the device.

In our display driver module, we provide four graphical primitives: boxes, polygons, line segments and text. We define procedures to encapsulate the details of driving the display.

```
procedure box(integer left,bottom,right,top, color);
procedure poly(integer array(1 to *) x,y; integer color);
procedure line(integer x1,y1, x2,y2, color);
procedure text(integer x,y; string s;  integer color);
```

Virtual Screen Coordinates

The coordinates we pass to those procedures refer to the locations on the physical device at which we wish to make the figure. We call these the *physical screen coordinates*. If we want to draw a box on a certain part of the display, we must know the mapping of screen coordinates onto the display. We can know this from the display's documentation, but we want to avoid rewriting our programs to run with different display devices. Different displays have their origins at different places on the display. So a dot at 0,0 might be in the middle of the screen on one display, but off the screen entirely on another. In addition, we envision our chip with the X dimension increasing to the right and Y increasing upward: a *left handed coordinate system*. Some devices have Y increase downward. If we plot on a display like that, the image will appear mirrored upside down.

We need a different coordinate system if we want to place things on specific parts of the display. Instead of drawing in physical screen coordinates, we draw in *virtual screen coordinates*. Virtual screen coordinates always increase to the left and up. The origin of virtual screen coordinates is always at the lower left of the screen. The procedures to draw on the screen may negate one coordinate and may have to add an offset to all coordinates to make it appear that 0,0 is at the lower left.

The pixels on a display may not be square. That is, a display may have a different number of pixels per inch in X than in Y. If we draw a square 30 by 30 pixels on such a display, it will not appear square, it will appear as a rectangle. To ensure that pixels in our virtual screen coordinates are square, we scale one dimension of our coordinates, if necessary, when we convert from virtual screen coordinates to the physical screen coordinates. So, besides negating a coordinate and adding an offset, our display driver module may multiply one coordinate by a scale factor to make squares come out square.

The procedures for drawing in virtual screen coordinates look the same as those for drawing directly in real screen coordinates, and they may be identical for some displays, but for the sake of device independence, we never have programs draw in real device coordinates, only virtual device coordinates.

```
procedure screenBox(integer left,bottom,right,top, color);
procedure screenPoly(integer array(1 to *) x,y;
   integer color);
procedure screenLine(integer x1,y1, x2,y2;  integer color)
procedure screenText(integer x,y; string s; integer color)
```

Assume that we have a display device that has its origin at the upper left. X coordinates increase to the right, Y coordinates increase downward. The display consists of 1024 pixels in each dimension. The display is 10 cm square. The mapping from virtual screen coordinates to real device coordinates is given below. No scaling is required because the ratio of the numbers of pixels in the X and Y dimensions is the same as the ratio of the physical dimensions of the screen.

$X_r = X_v$
$Y_r = 1024\text{-}Y_v$

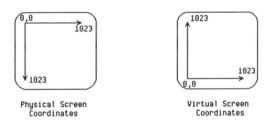

Physical Screen
Coordinates

Virtual Screen
Coordinates

Text books on graphics techniques define *normalized device coordinates* that range from -1 to +1 in one dimension and -aspectRatio to +aspectRatio in the other dimension. Like our virtual screen coordinates, normalized device coordinates are used for positioning in the display area. With our virtual screen coordinates, the left bottom corner is at 0,0, and we get top and right by calling a procedure in the graphics code that returns those numbers. To draw something near the top edge of the screen, we first call the procedure to find the virtual coordinate of the top edge. If we had used the textbook normalized device coordinates, we would have had to make a similar call to find the aspect ratio of the display to find the top edge. Normalized device coordinates require scaling in both dimensions. Virtual screen coordinates require scaling in only one dimension, so they are more efficient.

User Coordinates

Virtual screen coordinates allow us to draw anywhere on the screen, but we want to draw in a coordinate system that is relevant to our application. If we are drawing road maps, we want the screen to span miles, if we want to display a layout, we want the screen to span only a few lambda. We may also want to have our coordinates displaced as well, so 0,0 is not always at the lower left of the screen.

We define a third set of coordinates, *user coordinates*, that are scaled to the dimensions of our problem space. The conversion from user coordinates to virtual screen coordinates requires an offset and multiplication by a scale factor. As we shall see, user coordinates may also be mirrored or rotated from the virtual device coordinates. We supply additional procedures for drawing in user coordinates:

```
procedure drawBox(real left,bottom,right,top, color);
procedure drawPoly(real array(1 to *) x,y;
   integer color);
procedure drawLine(real x1,y1, x2,y2; integer color);
procedure drawText(real x,y; string s; integer color);
```

Clipping and Viewports

Many displays draw unpredictable figures if we try to draw outside the legal area of the display. Our graphics package prevents this by *clipping* the figures to the display area. We only display those lines and parts of lines that are in the display area.

We can use the same clipping feature to limit our drawing to one area of the screen, called a *viewport*. Initially, we set the viewport to cover the entire display area, but we can set a new viewport simply by setting a new clipping rectangle. We set the viewport in virtual screen coordinates, because it refers to an area of the screen. However, we can clip using user coordinates or virtual screen coordinates. Clipping in user coordinates is more efficient since is saves translating clipped lines into screen coordinates.

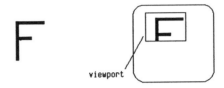

The Cohen-Sutherland Clipping Algorithm

This section describes a segment-clipping algorithm. We use the segment clipping algorithm to determine which part of a line to draw on the screen and also to determine which portions of polygons are visible on the screen. Rectangles are easy to clip and should be treated specially. This algorithm consists of a quick test to eliminate segments that are "obviously" off the screen, followed by an algorithm to determine the part of the segment that is on the screen and return it.

First, we imagine the space of user coordinates cut by the four edges of the clip box. Each region has a four-bit *point code* that represents its position with respect to each of the edges of the window. The point code is assigned as follows:

1001	1000	1010
0001	0000 (screen)	0010
0101	0100	0110

The leftmost bit indicates a point above the line, the next bit below, followed by right and left. No bits means the point is inside the visible screen area.

If both point codes are 0000, both points are in the visible screen area, so the whole segment is visible on the screen, and we draw the whole segment. If the logical AND of the two codes is non-zero, the segment is entirely to one side of the screen, and we may safely skip the segment without further processing. Otherwise, part of the line may be visible. To determine if it is visible, we clip one point that is beyond an edge to that edge, recalculate the point code and try again. One of the above tests will work within four iterations.

For example, in the following figure, line segment A has endpoints with point codes 1001 and 0101. Since the AND is non-zero, the line is off the screen. Line segments B and C both may be on the screen, according to the algorithm. Clipping eventually shows line segment B to have a short visible piece and line segment C to be off the screen entirely.

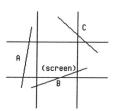

The following code should help explain the clipping algorithm. The procedure clipSegment returns a boolean telling whether or not the segment is visible in the rectangle bounded by left, bottom, right, and top. If some part of the segment is visible, clipSegment modifies the values of the input variables to clip the endpoints of the segments to the edges of the window. A full treatment of this clipping algorithm and options can be found in Newman and Sproull (1979)..

```
boolean procedure clipSegment(
  modifies real x1,y1,x2,y2;
);
begin
  bits c1,c2;
  c1 := pointCode(x1,y1);
  c2 := pointCode(x2,y2);
  while not (c1=c2='b0) do begin
    if (c1 msk c2)<>'0 then return(false);    # not visible
    if c1='b0 then begin         # exchange points 1 and 2
      exchange x1,y1,c1 with x2,y2,c2
    end;
    if (c1 msk 'b1)<>'0 then begin              # clip left
      y1 := y1 + ((y2-y1) * (left-x1) / (x2-x1));
      x1 := left;
    end else if (c1 msk 'b10)<>'0 then begin   # clip right
      y1 := y1 + ((y2-y1) * (right-x1) / (x2-x1));
      x1 := right;
    end else if (c1 msk 'b100)<>'0 then begin  # clip bottom
      x1 := x1 + ((x2-x1) * (bottom-y1) / (y2-y1));
      y1 := bottom;
    end else if (c1 msk 'b1000)<>'0 then begin # clip top
      x1 := x1 + ((x2-x1) * (top-y1) / (y2-y1));
```

```
      y1 := top;
    end;
    c1 := pointCode(x1,y1);   # code for new x1,y1
  end;
  return(true);
end;
```

Transformations

We draw instances of cells that are *transformed*, mirrored and rotated. Instead of transforming our coordinates in the application program, we set up the transformation in the graphics package. We send our data to the display as if they were not transformed. The graphics package transforms them and clips the transformed features. When we are finished drawing a transformed figure, we remove the transformation from our user coordinates.

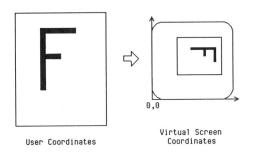

User Coordinates   Virtual Screen Coordinates

Kinds of Transformations

The most common transformation is *translation*. A cell is to be displayed offset from its defined location. The transformation of coordinates is:

$x' = x+dx$
$y' = y+dy$

To *rotate* counterclockwise by an angle θ:

$x' = x\cos\theta - y\sin\theta$
$y' = x\sin\theta + y\cos\theta$

We display the same features scaled by multiplying the coordinates by a *scale factor*:

Graphics

$x' = xS_x$
$y' = yS_y$

Mirroring is a special case of scaling in which we negate one coordinate. To mirror across the Y axis, we negate the x coordinate:

$x' = -x$
$y' = y$

Concatenating Transformations

We may have nested calls to transformed subcells. In addition, we have a transformation from user coordinates to virtual screen coordinates and from virtual screen coordinates to real screen coordinates. For efficiency, we can concatenate multiple transforms into one transform. So between user coordinates and real device coordinates, we have only one transformation, no matter how many levels down the hierarchy we are displaying and no matter how great the disparity between virtual device coordinates and real device coordinates. For example to mirror across the Y axis, then translate by (10,20):

$x' = -x$
$y' = y$
$x'' = x'+10 = 10-x$
$y'' = y'+20 = y+20$

If we are using our sample display from page 46, the mapping from virtual screen coordinates to real screen coordinates is:

$x''' = x''$
$y''' = 1024-y''$

So the the resulting transformation to real screen coordinates is:

$x''' = 10-x$
$y''' = 1004-y$

The order in which we concatenate transformations is important. Mirroring in X then rotating 90 degrees counterclockwise yields the transformation:

$x' = -y$
$y' = -x$

F ⇨ ㄱ ⇨ ⌐

Normal   MX   MX Rot

Taken in the other order, the transformations concatenate to produce:

$x' = y$
$y' = x$

F ⇨ ⌊ ⇨ ⌋

Normal   Rot   Rot MX

Matrix Representation of Transforms

The coordinate pair can be viewed as a vector: $[x\ y\ 1]$ and the transformation can be viewed as a matrix:

$$\begin{bmatrix} S_x \cos\theta & S_x \sin\theta & 0 \\ -S_y \sin\theta & S_y \cos\theta & 0 \\ dx & dy & 1 \end{bmatrix}$$

where S_x and S_y are the scale factors in the x and y dimensions including mirroring, θ is the counterclockwise rotation angle, and dx and dy are the translation parts of the transformation.

$$[x'\ y'\ 1] = [x\ y\ 1] \begin{bmatrix} S_x \cos\theta & S_x \sin\theta & 0 \\ -S_y \sin\theta & S_y \cos\theta & 0 \\ dx & dy & 1 \end{bmatrix}$$

Multiple transformations can be concatenated as matrix multiplications into one transform. The sequence of transformations can be read left to right in the concatenation of transforms. For a mirror followed by a rotation, the transforms are concatenated as: $[Tr] = [Mir][Rot]$.

Simplifications

The matrix formulation of the transformation makes concatenation of transforms obvious, but we can simplify the calculation of the transformation by dealing the the transformation as equations. The equations for transforming coordinates pairs are:

$x' = xS_x\cos\theta - yS_y\sin\theta + dx$
$y' = xS_x\sin\theta + yS_y\cos\theta + dy$

These equations require two trigonometric calculations and eight multiplications. In our implementation, we use scaling for three purposes: to ensure that pixels are square in virtual screen coordinates, to choose a magnification in user coordinates and to mirror our data. The first two uses can be separated out of the transformation and put into the graphics package as a multiplication after all other transformations have been done. We do not scale parts of our data differently than others, so elimination of scaling from the equations does not restrict our implementation. The remaining mirroring is either +1 or -1 so it doesn't require multiplication. We save four multiplications.

$x' = xM_x\cos\theta - yM_x\sin\theta + dx$
$y' = xM_y\sin\theta + yM_y\cos\theta + dy$

In our integrated circuit designs, we limit our rotations to be multiples of 90 degrees, so the entire combination of mirrors, sines and cosines can produce only -1, 0 or +1, which we encode in the orientation number described in chapter 2.

We now have three numbers describing the transformation: the orientation number, dx and dy. These simplifications do not unduly restrict layouts, but simplify our code tremendously, since we can now mirror and rotate a point with a case statement on the orientation number. We can implement concatenation of transforms with a table lookup and a case statement to combine the translations correctly.

```
procedure transformPoint(integer orient;   real dx,dy;
   real x,y;   produces real xout,yout);
begin
   case orient of begin
      [transIdentity] begin xp:=x;    yp:=y   end;
      [transRot1]     begin xp:=-y;   yp:=x   end;
      [transRot2]     begin xp:=-x;   yp:=-y  end;
      [transRot3]     begin xp:=y;    yp:=-x  end;
      [transMX]       begin xp:=-x;   yp:=y   end;
      [transMXRot1]   begin xp:=-y;   yp:=-x  end;
      [transMXRot2]   begin xp:=x;    yp:=-y  end;
      [transMXRot3]   begin xp:=y;    yp:=x   end;
   end;
   xout := xp + dx;
   yout := yp + dy;
end;
```

Transforms

We have already seen that we must transform coordinates when calculating bounding boxes of instances. We implement a separate module for transformations that includes procedures to concatenate transforms, to transform points and boxes and to inverse transform points and boxes.

Transforming User Coordinates

In the graphics package, we use the transformation procedures to set the transformation on user coordinates. The most commonly-used procedures for manipulating transforms in the graphics package are pushTransform and popTransform. pushTransform adds a new transform to the *front* of the current transform, pushing the current transform onto a stack. popTransform pops the last transform off the stack of transforms. We use these procedures when we wish to display a sub-cell of a chip mirrored or rotated. That sub-cell may have a sub-sub-cell that is further mirrored and translated. When we have finished plotting the cell, we call popTransform to restore the previous transform.

Using these procedures, we display a cell rotated by 90 degrees counterclockwise by:
```
grafix.pushTransform(transRot90,0.L,0.L);
drawCell(cellName);
grafix.popTransform;
```

Windows

Since transformations are expensive, we want to clip before transforming. To do this, we reverse transform the sides of the viewport and implement our clipping algorithm using user coordinates. This clipping rectangle in user coordinates represents the viewport and is called the *window*.

This definition of window is consistent with the definition given in graphics texts. Display-oriented editors and CAD software usually use the term *window* to refer to an area on the screen, that which we call a viewport. We will use the graphics definition.

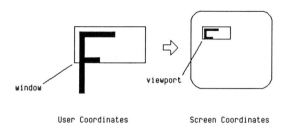

A Note About Text

Although we transform the location at which we place text on the display, we always draw the text itself left-to-right, right-side up, unmirrored. Most graphics devices provide text in this fashion as a built-in feature, so it is fast and easy to implement. In addition, text in its normal orientation is easier to read. One of the assignments deals with alternative implementations of text.

Scaling Procedures

The scale adjustment of coordinates in our graphics package consists of two parts: a device-dependent aspect ratio correction, which we use to convert from virtual device coordinates to real device coordinates, and a single scale factor, which we use for zooming in and out on the display in user coordinates. All code using the display package sees only one scale number for zooming in and out in user coordinates. We do not scale X and Y dimensions separately, but since integrated circuits are physical objects, this does not cause a problem with our application. We multiply all coordinates by the scale factor after applying the transformation from user to virtual screen coordinates.

Coordinate Conversion

As part of the graphics package, we include procedures to convert points from user coordinates to virtual screen coordinates and vice versa. We also have procedures to determine aspects of the physical device we are driving, such as the size of the screen and the size of text in virtual screen coordinates. Coordinate conversions take into account all scaling and transformations.

Summary of Graphics Output Features

The graphics module provides three functions for graphics output:

- **mapping** -- translating coordinates from the problem domain into the coordinates of the display device. During mapping, we perform mirroring, rotation and scaling.

- **clipping** -- filtering data so only data in a restricted area on the screen is actually sent to the display device. This facility allows us to show our data in a limited part of the screen, called a *viewport*.

- **driving** -- translating drawing commands into a form that is acceptable to the display device and sending that data to the display device.

Summary of Coordinate Systems

The graphics package deals with three different coordinate sets: *user coordinates*, *virtual screen coordinates*, and *physical device coordinates*. It includes facilities to plot using user and virtual screen coordinates and to transform coordinates for client programs.

User Coordinates

User coordinates represent the user's data space. Since we deal in lambda on a chip, we set the user coordinates to lambda. The display module transforms those coordinates to the device. The coordinate space is left-handed, with x increasing to the left and y increasing upward. Negative numbers are allowed.

Virtual Screen Coordinates

Virtual screen coordinates represent the screen. Virtual screen coordinates have their origin 0,0 at the lower left corner of the display. The graphics module has a procedure to return the virtual screen coordinates of the upper right corner. Pixels in virtual screen coordinates are square. Virtual screen coordinates are the only screen coordinates that programs use. They are necessary for users when putting features on a specific part of the screen.

Physical Device Coordinates

Physical device coordinates are the coordinates that the device recognizes. They vary from one display to the next. There may be different numbers of pixels and the origin may be at the top, bottom or middle. We deal with physical device coordinates in only one place -- the display driver module. All application programs deal with virtual screen coordinates and user coordinates.

Color

We use colors to distinguish layers of an integrated circuit. We are interested not only in the color of single layers, but also with interactions between layers. The need for large numbers of distinguishable colors creates problems.

This section describes three different color implementation strategies. Successful systems, such as Caesar (Ousterhout 1983) have been built using combinations of all three different strategies. They are described in order of their difficulty of implementation.

Transparent Colors

We would like to view an integrated circuit layout as several layers superimposed on the screen. We want to be able to see every layer and, when two layers overlap, we would like to see the two colors merge so we can see that both layers are present in the overlap area.

We can implement this feature by dedicating one bit in every pixel to each layer. We call the one bit per pixel a *bit plane*. We draw on a layer by drawing on the bit plane for that layer. When two layers overlap, the pixels in the overlap contain the logical OR of the bits of the two colors.

We set up the color map in the display so the overlap color looks like the overlap of translucent colors. For example, we can draw a rectangle in color 1 over a color 0 background, as shown on the left. If we then overlap it with a rectangle in color 2, we get color 3 at the interaction, shown at the right.

```
0 0 0 0 0 0 0 0 0 0          0 0 0 0 0 0 0 0 0 0
0 1 1 1 1 1 1 1 1 0          0 1 1 1 1 1 1 1 1 0
0 1 1 1 1 1 1 1 1 0          0 1 1 1 1 1 1 1 1 0
0 1 1 1 1 1 1 1 1 0          0 1 1 3 3 3 1 1 1 0
0 1 1 1 1 1 1 1 1 0          0 1 1 3 3 3 1 1 1 0
0 0 0 0 0 0 0 0 0 0          0 0 0 2 2 2 0 0 0 0
0 0 0 0 0 0 0 0 0 0          0 0 0 2 2 2 0 0 0 0
0 0 0 0 0 0 0 0 0 0          0 0 0 2 2 2 0 0 0 0
0 0 0 0 0 0 0 0 0 0          0 0 0 0 0 0 0 0 0 0
```

If we set the color map so color 1 is yellow, color 2 is blue and color 3 is green, a user will see a yellow box and a blue box overlapping and showing green in the intersection.

Most raster display devices implement this feature with the ability to logical-OR bits into the pixels of the bitmap when drawing. We can implement transparent colors for all layers using one bit per layer. A complex manufacturing process may have twenty layers, requiring twenty bits per pixel. Such display devices do exist, but they are expensive. Further, human eyes are incapable of distinguishing the two million different colors.

This method of implementing colors is very simple, and may be suitable for applications where the number of layers needed in a design are limited. In addition, we may be able to re-use colors or whole bit planes by assigning more than one layer to a bit plane. We then rely on a user to distinguish the difference by context. There are other ways to approach this problem as well: opaque colors and stipple patterns.

Opaque Colors

We can define some of our colors as *opaque colors*. Opaque colors replace the existing colors in the display. The screen illusion is that they lie on top of everything else. When we draw in an opaque color, a user can not see the layer interactions, but there are many occasions when it is not important to see the interactions, such as when displaying text and drawing contact cuts.

Opaque colors require only one color number in the pixel, rather than one dedicated bit in the pixel. Therefore, we can have many more of them: an eight-bit display can have 256 opaque colors, but only 8 transparent colors.

Stipple Patterns

Where we have very few bits per pixel, we don't even have enough colors to implement opaque colors for all layers. The most extreme such case occurs on a black and white display. Since there is only one bit per pixel, we have only two colors: black and white.

To distinguish layers, we use a technique known as *stippling*, or *dithering*, in which we draw a different pattern for each layer, instead of drawing the solid color. Trading resolution for shading, we replicate a square bit pattern over the entire area to be covered. Only those pixels that correspond to a 1 in the stipple pattern are written to the display. Solid colors have all bits in the stipple set to 1, and we define colors with different density by choosing different stipple patterns.

For example, given a 4 by 4 stipple pattern, we can draw areas shaded with vertical bars with any of these patterns:

All four of these patterns generate the same pattern on the display so a user wouldn't be able to identify different layers drawn with these stipples. When defining stipple patterns, avoid patterns that differ by a shift.

When defining stipple patterns, we take into account not only the pattern of dots, but also the density of the dots. A short distance from the screen, the stipple patterns is too fine to see, so a user sees only the gray level of the density of the pattern. Two stipples with different patterns, but the same number of dots turned-on become indistinguishable. This problem is greater for higher resolution displays because the pixels are smaller and for hardcopy plots because they are much higher resolution and are often viewed at greater distance.

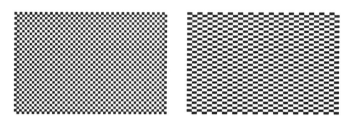

Stippling is used with both black and white displays and color, particularly with hardcopy color plotters that have high density but only one bit per pixel. Stippling works with both opaque and transparent colors.

One commonly-used method to implement colors on a display is to use sparse stipple patterns with color for all layers. Layer-to-layer interactions are visible from both the pattern and color overlap. In such a system, the edges of the graphical items may not be readily apparent, so the graphical items may be outlined to emphasize the borders. Outlining slows down the display, since more data must be drawn, and some method must be provided to remove interior lines on a layer.

Color Numbers

Combining transparent, opaque and stippled color, we can define a color to have three parts: a color on the display, a bit which defines whether or not it is opaque, and a stipple pattern. The color chooses the value we write into the pixels on the display. The opaqueness chooses the operation we use to merge the new values into the old colors already on the display, OR for transparent, REPLACE for opaque. The stipple pattern determines which pixels in the area are actually changed.

There is no need to deal with color handling outside our plotting software. We assign color numbers that refer to an index containing the screen color value, the transparency boolean and stipple pattern. The plotting software looks up the color to determine what to do with it, so our color numbers are device independent. For each display device, though, we define the mapping from the integers we use for colors to the triple of true color, transparency and stipple.

Graphical Input

Graphical input is easier to implement than graphical output because there are very few options. We are concerned with combining keystrokes with cursor positions and transitions on buttons on the pointing device.

There are two common pointing devices used in CAD software, a *tablet* and a *mouse*. A tablet senses the position of a movable puck over its surface; a mouse senses its motion over a table top. Both devices have one or more buttons on the top of the movable part. A user moves the mouse or tablet puck to move the cursor on the display, then presses a button on the mouse or tablet or a character on the keyboard to invoke a command.

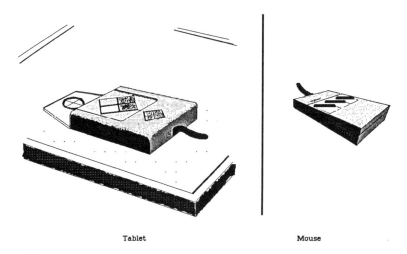

Tablet   Mouse

From our point of view inside the program, we must keep the cursor on the screen updated as the pointing device moves and we must supply coordinates of the cursor and positions of the buttons on the pointing device to the application program. Updating the cursor must happen very rapidly and most graphic devices have a mechanism to tie the pointing device directly to the display device. We supply the input and translation ourselves.

```
procedure userInput(
   produces boolean pointingDeviceEvent;
   produces integer value,  sx,sy);
```

This procedure returns input from the pointing device and keyboard. If `pointingDeviceEvent` is true, then `value` indicates which buttons are pressed on the pointing device and `sx` and `sy` are the virtual screen coordinates of the cursor. If this is not a pointing event, `value` is the character value of the key that was typed and the contents of `sx` and `sy` are not defined.

This mechanism merges pointing and character input into a single procedure call. Some graphics terminals merge the button presses and cursor positions as escape sequences in the character stream from the display. We still must process the stream to convert coordinates from real device coordinates to virtual screen coordinates.

Menus

A *menu* is a list of words or pictures on the screen. Each word or picture in the menu is called a *menu entry*.

The menu resides in a dedicated space in virtual screen coordinates. That space is usually along one edge of the screen. We set up the menu by setting a new clipping box for the menu area and we draw the menu entries in the menu area. When we handle a button press on the pointing device, we first check to see if the user pointed inside the menu area. If so, we determine which command the cursor was nearest and return the code for that command. The code could be an integer code or the string of a command name.

The menu facility should be a general one. There may be more than one menu on the screen at a time. There may be commands that change the menu entries, so the menu would have to be redrawn. There is no fundamental reason why the menu must be part of the graphics package, it may be a separate module that the user invokes for every button press on the pointing device. However, the new viewport must be passed to the graphics package, so the other information is not drawn over the menu.

Dynamic Displays

CAD software, particularly on personal workstations, devotes large amounts of time to running the display. Since the processor is waiting for input at human speeds, we can do a lot of processing to give interactive feedback during user interaction. This feedback takes the form that we will call *dynamic displays*.

A *dynamic display item* is a figure on the screen that appears for a short time as an aid or guide for user input. Dynamic displays are generally graphical in nature, but may include text. CAD software uses several kinds of dynamic displays:

- An *alignment mark* is an indicator on the screen used as a point of reference in a graphical editor. The cross in this figure is the alignment mark. The arrow is the cursor that tracks the pointing device. When a user presses a button on the pointing device, the alignment mark moves to the new position. Alignment marks may be rectangles or lines as well.

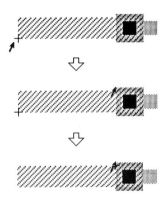

- A *rubber band line* is a line with one point fixed and one that follows the cursor. A *rubber band box* is a rectangle with one corner that follows the cursor. Rubber band objects change size as the user moves the cursor around the screen. In this figure, the rubber band box starts at the first point and follows the cursor through the second and third steps.

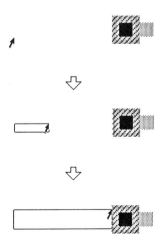

- *dragging* is an operation where a piece of the drawing follows the cursor around the screen. In this figure, the user is dragging the box up and to the left of the contact.

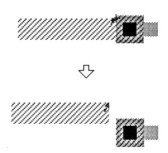

- A *popup menu* is a menu of commands that appears when a user presses a button on the pointing device. The menu appears under the cursor, lying on top of the drawing. When the user selects an entry, the menu vanishes. A popup menu may appear under the cursor, as shown in the following figure, or it may appear only when a user points to a sensitive menu spot on the screen.

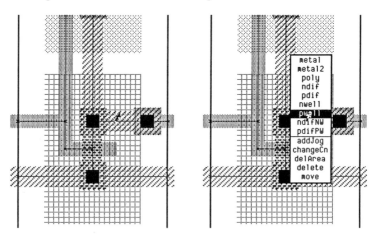

Dynamic displays happen quickly, some at the speed with which the cursor tracks the pointing device. All dynamic displays can be easily implemented using a display-list graphics device. A display-list device keeps a list of items to be displayed and re-displays them every screen refresh. They are typically limited in the amount they can display, but the display list can be quickly modified to change the position of the items in it and the items can be quickly removed when the display is finished.

The easiest way to implement a dynamic display on a raster device is by exclusive-ORing bits onto the screen. The bits can be removed later by exclusive-ORing again. This technique is simple, but it does have problems. If we draw a large, complex item over varying colors, it is hard to see because of the varying colors. For this reason the exclusive-OR method is restricted to simple items and black and white displays. It is frequently used for small alignment marks and rubber-band boxes.

Many raster displays have off-screen display memory that can be effectively used for dragging as well as for the simpler items. We can copy an area from the screen to off-screen display memory quickly and write the dynamic display item on the screen. When we are done with the display, we copy the area back. We implement dragging by continually saving, drawing, restoring and re-drawing the moved data.

For displays without off-screen memory, we can dedicate two bits from each pixel in the display to dynamic displays. We set up the color map so that 10 in those pixel locations is always white, no matter what the other pixels are, and 11 in those pixel locations is always black. If the first bit in the pixel is 0 then the other bits in the pixel determine the color of the pixel.

We draw a dynamic display by setting those two bits to make the pixel white or black. We erase by clearing the first bit. This technique requires two precious bit planes from the display. It decreases the number of colors we can display by a factor of four regardless of the number of bits per pixel. But this may be our only way to build dynamic displays and we have other ways to differentiate colors.

Exercises

Programming Problems

1. Write a graphics package for a display device you have. Include virtual screen coordinates, user coordinates, clipping and transformations. Draw boxes and text. If you have no graphics display, write a driver for a text terminal. How do you implement color on a text terminal? How much of your device driver is useful for some other device?

2. Implement the following names for color assignments on your display. You may choose any implementation style you wish.

0. black	4. yellow
1. blue	5. cyan (or pale blue)
2. red	6. magenta (or pale red)
3. green	7. white

3. Write a menu handler for a menu of words on the right side of the screen. Your menu handler must draw the menu and handle pointing into the menu area. It should return the text of the word that was selected.

4. Implement code to save user input in a file and to take input from a replay file. Should the code be part of the graphics software? Why or why not?

5. The Cohen Sutherland clipping algorithm works for line segments at all angles. Modify the algorithm to handle the common case of horizontal and vertical lines quickly. Write a procedure for clipping boxes and polygons.

Questions

1. Refer to the sample device on page 46. What is the translation from virtual device coordinates to real device coordinates if the screen is 10cm by 5cm with 1024 x 512 pixels? 10 x 10 cm with 1024 x 768 pixels?

2. How do you clip text? What do you do when only half a text character fits on the screen? Describe a method for handling text that can draw the partial characters.

3. Refer to the matrix description of the full transformation. Is the translation by **dx,dy** done before the mirroring and rotation or after?

4. What changes would you have to make to drawing procedures if text can be rotated to any angle? How is the problem simplified because we only use the 8 orthogonal transformations?

5. Using 64-bit long real numbers, how large an area do you estimate you can represent in user coordinates? Is this reasonable?

6. What problems do you forsee from using real numbers? What solutions do you propose?

7. What are the impediments to using 32-bit integers for user coordinates instead of 64-bit reals?

8. Consider menus implemented as part of the graphics package. A menu consists of a column of words on the right side of the screen. When a user clicks in the menu area, the menu processing code determines the menu item to which he pointed. The call to get the next user input returns not only characters and button presses, but a menu item identifier when a menu is invoked. How would you implement this change to the graphics package? What are the advantages and disadvantages of this interface over a menu implemented separately?

References

J.D. Foley and A. Van Dam, *Principles of Interactive Computer Graphics*, Addison-Wesley, 1982.

J.K. Ousterhout, "The User Interface and Implementation of an IC Layout Editor", *Transactions on Computer-Aided Design of Integrated Circuits and Systems*, vol. CAD-3, No. 3, July, 1984.

W.M. Newman and R.F. Sproull, *Principles of Interactive Computer Graphics, Second Edition* McGraw Hill Book Company, 1979.

CHAPTER 4

PLOTTING LAYOUT

Plotting an integrated circuit design serves a dual purpose. First, it is a means for documenting a design. A hardcopy plot of the layout implicitly carries with it all the information about the design. Secondly, plotting is the time-honored method of verifying the correctness of a design. As a representation of the manufactured product, it is a preferred form for verification. In the absence of automated checking tools, plotting and verification by eye is a workable, though imprecise, technique for verifying designs.

In this chapter we develop a plotting program, a program that reads the layout files and displays their contents graphically on a hardcopy plotter or softcopy display. The plotting program, also called a *plotter* uses the parser we developed in Chapter 2 and the graphics procedures from Chapter 3. We will use the plotting program as a stepping stone to the layout editor in Chapter 5.

The plotting program can be divided into four major pieces: a parser for layout files; a semantics module to build the plotter data structures; the graphics module for displaying the data; and a command loop used to interact with a user.

The Core of a Plotter

We begin our examination of plotters by discussing the basic plotting engine: a command loop and plotting interface. After this discussion, we will consider additional features and options for plotting. Later sections deal with the difficulties imposed by physical properties of output devices.

Use of the plotter follows a rather simple flow pattern. A user loads a layout file into internal storage, sets plotting options, and plots the data on the output device. He may wish to plot several different cells, plot at different scales, or plot different parts of the same cell. Our code must be able to handle these options.

The Command Loop

The flow we just described suggests a program that queries the user for each piece of data in turn: the file to load, the area to plot, the output device, and so on. Once the last piece of data is available, the program generates the plot and terminates.

Such an organization is easy to program, but it is not easy to use. A user must give the relevant data in the order your program wishes to consume it, which may not be the order in which it is organized. In addition, the organization is inflexible. Suppose a user wants to plot several different areas of the same cell. He would have to re-load the cell each time and re-set all parameters. We use a different interface, a command driven system, rather than the query-driven approach.

The plotter we develop is *command driven*, it accepts input as a sequence of commands rather than prompting for input. When plotting a file, a user gives commands to read layout files, set options and generate plots. These commands can be collected into *command files* to drive the plotter in a batch-like mode.

The main interaction is in the `handleCommands` procedure:

```
procedure handleCommands(pointer(textFile) f);
begin
  string command;
  setInputFile(f);
  while not eof(f) do begin
    command := getString;
    if command="read" then begin
      # take user commands from a file
      open(ff,getString,input);
      if ff then begin
        handleCommands(ff);
        close(ff);
        setInputFile(f);
      end;
    end else if command="plot" then begin
      currentCell := lookupCell(getString);
      if currentCell=nullPointer
```

Plotting Layout

```
        then error("Cell not found")
        else drawCell(currentCell);
    end else if command="quit" then begin
      return;
    end else if command="load" then begin
      ..
    end else if ...

    ...

    end else begin
      error("command "&command&" not recognized.");
    end; # command selection
  end; # while not eof loop
end;
```

getString gets a string from the current input file, set with setInputFile. The command read opens a new input file and calls handleCommands recursively to accept commands from a file. quit ends the procedure, ending the command file or the plotting session.

This general form is useful in any interactive program. There are some obvious improvements that you can make, such as removal of the case sensitivity, or acceptance of any unique prefix as a legal command identifier.

Loading a File

To load a file we call the parse procedure in the layoutParser with the name of the file to read and the semantics, then use the cells list as our data structure. The plotter does no searching or sorting, so the list data structures from Chapter 2 are adequate.

```
lysem := new(lySemantics);
lypars := new(layoutParser);
fileName := getString;
lypars.parse(fileName,lysem);   # do the work
cells := lysem.cells;           # get the cells
```

Displaying a Cell

The plot command takes the name of the cell, looks up the cell in the list of cells and plots it by calling procedures in the graphics package. The procedures in the graphics package correspond to the kinds of items we wish to display, so the job of drawing a cell is rather simple. The main drawing procedure is given here:

```
procedure drawCell(pointer(cellClass) c);
begin
  for all c.data (d) do begin
    # first check that part of the item is visible
    if boxesIntersect(d.bb,grafix.window) then begin
      case d.type of begin
        [boxType] begin
          grafix.drawBox(...);
        end;
        [polyType] begin
          grafix.drawPoly(...);
        end;
        [connType] begin
          grafix.drawBox(...);
          grafix.drawText(d.text);
        end;
        [instType] begin
          grafix.pushTransform(inst.orient,inst.x,inst.y);
          drawCell(inst.cell);
          grafix.popTransform;
        end;
      end;
    end;
  end;
end;
```

The procedure `drawCell` is recursive. When we find an instance, we push its transform and call `drawCell` again. The check that each item is visible is an optimization. The procedure would work properly without it because items being plotted outside the viewport would be clipped in the graphics package. But the bounding box check eliminates from consideration instances and polygons that fall outside the displayed area. The performance improvement can be substantial when plotting a small area of a chip.

Necessary Additions

We have covered the fundamentals of plotting, the plotting engine. We now describe some of the features and options required for a successful plotting program. To make a minimally-acceptable system, we allow users to control the portion of the data that will be displayed, view the cells that have been read and select the output device.

Window. We set the region of the cell we wish to plot by giving the window in user coordinates. The window command sets both the scale and the translation of an initial transform. The default window is the bounding box of the cell.

Viewport. The default viewport is the full size of the plotting device. We may provide an option to plot the cell on a region of the output device. We specify the region in virtual screen coordinates.

Plot Selected Layers. We can set the plotter to only plot selected layers of the cell. This feature is essential when plotting cells designed in complex processes and for checking design rules by eye.

Multi-Sheet Plots. A user may wish to make a plot of a cell that is larger than the size of the plotting device. It is possible to do this manually, setting the window to each different area to plot, but it is common enough with hardcopy plotters that we can make a command to set the scale of the plot independent of the window, and have the plotter calculate the proper window coordinates for each of several sheets to be plotted.

View Hierarchy Level. Plotting the details of all cells in the hierarchy may not be necessary in all cases. Omitting low levels of hierarchy reduces plotting time. If a user is concerned with only one or two levels of the hierarchy, he may set the `level` of the plotter to omit the lower levels, plotting them as outlines with the cell name centered in it. This feature may be extended to allow the omission of individual cells or instances of cells.

Set Device. The plotting software may output to more than one kind of device. Users may wish to plot on the display to quickly find the proper window for plotting on a slower plotter.

Connectors On/Off. Users may wish to view the plot with or without connectors.

Text On/Off. Names of connectors may clutter the screen, especially on large-scale plots. Users should be able to turn off the text, so only the graphics is plotted.

List Cells. We print a list of the cells in memory with their bounding box sizes to indicate which cells are available for plotting.

Minimum Feature Size. When a feature is too small to be seen or is too small to be resolved on the plotter, it can safely be omitted from the plot. With this feature, large-scale plots run much faster. This minimum feature size is normally a function of the plotting device, but it is useful for a user to set it manually if small-scale features are of no interest.

Help. Every interactive program must have a minimum of on-line documentation. This documentation need only be the program identification, version number and a list of the commands with one-line descriptions. Usually, this feature is used to find the correct command name that a user already knows is present, rather than as a reference for learning about the tool.

Desirable Features

Grid On/Off, Size. Some users like to see a grid on the plot to provide a reference measurement. The grid may consist of lines or dots at the grid points, or a combination.

Plot Selected Sheets. This command allows a user to select only a few sheets of a multi-sheet plot for plotting.

Initial Transform. This command sets an initial transformation to be applied to the cell before it is plotted. It provides an easy way to plot a cell rotated, for example, so it fits better on the page.

Initialization File. Many users want to set their own mapping from layers in their chip to colors on the screen. To provide this, the plotter should initialize that mapping from an initialization file. Individual users can customize the file as they wish. When a user starts a program, we read the initialization file to set internal color map tables.

Practical Considerations

Most plots are generated to obtain hardcopy. Hardcopy plotters are physical devices with peculiarities due to their implementation. These peculiarities affect the way we present the data to the plotter both to get the plotter to accept the data and to use the plotting device efficiently.

Considerations for Pen Plotters

A pen plotter spends time to select the pen for the drawing color, to move to the beginning of a line to draw and to draw the line. The amount of time for each operation depends on the plotter. Changing pens is usually more expensive than pen movement and can take a second or more, so we want to do it infrequently. Typically, pen plotters make one pass through the data for each color, to reduce time spent changing pens.

Sorting For Pen Movement

To draw a line on the page, the plotter moves a pen to the beginning of the line, lowers the pen and draws to the other end of the line. We reduce plotting time by reducing this pen motion. The movement of the pen to actually draw cannot be eliminated, but movement of the pen from one line segment to the next can be reduced by properly sorting the edges.

This sorting operation has been shown to be NP-complete, it has no efficient solution. However, we don't really care to find the minimum, just a good sort, so our problem can be much simpler. The following are some sorting methods that work reasonably well. The discussions talk about edges to plot, but the arguments apply equally well to whole boxes, polygons, or instances. Recall that the goal is to reduce plotting time. If the optimization of the plot takes a significant amount of time, we gain nothing.

- **Natural Clustering.** Many programs rely on the "natural" clustering given by the hierarchy of the layout. The plotting algorithm we saw will plot all the data in an instance before returning to the containing cell. This method is simple, and often quite good, but its performance is unpredictable since it relies on the data being properly ordered by the way the user drew it. The performance can be very bad, especially with machine-generated layout.

- **Buffered Optimization.** Instead of plotting every edge as we find it in the data, we maintain a buffer of 100 edges to be plotted. We fill the buffer with edges and plot the edge that requires the least pen movement. Every time we plot an edge, we add a new edge to the buffer. When we reach the end of the data to be plotted, we flush the remaining edges from the buffer, plotting them.

  This algorithm is also easy to implement, but the 100 distance calculations per edge can be a significant overhead. In addition it has rather unpredictable behavior since the buffer can fill up with "bad" edges, reducing the effectiveness of the sort.

- **One-Dimensional Sort.** An obvious solution is to collect all the edges to be plotted, sort them in X or Y and plot them in sorted order. This solution looks attractive because sorting is a problem of known complexity and the result is reasonably good. However, in the worst case, this sorting produces a very bad sort. For example, suppose the three edges in the following figure are sorted by minimum X coordinate. They are plotted in order A, B, C. The order A, C, B would have been much better, since we could have plotted C from right to left and not had to move all the way back to the left for B.

- **Bin Partitioning.** In practice, full sorting is counterproductive. It is computationally expensive and produces a poor result. A straightforward partitioning into bins usually produces a better sort much faster than does a full sort in one dimension. We divide the cell area into bins and put each edge into the bin that corresponds to its part of the chip. We plot all edges in a bin before proceeding to the next. The number and size of the bins depends on the number of edges to plot.

Interior Edge Removal

When plotting two rectangles that touch on the same layer, we want to eliminate the line between them that is an artifact of the way they are represented in the data. Removing the edge makes a more readable plot and saves pen motion. Elimination of internal edges requires that we consider all edges in the plot before plotting, then only plot those that are "outside" edges. Merging polygons is a problem that we encounter again in design rule checking and circuit extraction. The polygon merge algorithm in Chapter 8 can be used to eliminate interior edges for a pen plotter.

Considerations for Raster Plotters

Rasterizing

Raster hardcopy plotters require that we send the data as a raster in the order it is to be plotted. Although some raster plotters are equipped with hardware and software to convert polygon edges to a raster, many devices require that we convert the data to the raster for them and order it for the plotting direction.

One simple solution to this problem is to allocate a raster array of bits the size we will need for the whole plot and define a plotting module that sets bits in our huge raster instead of interfacing directly with the device. We write different stipple patterns for different layers directly in the array. After all the data has been plotted into our array, we write it to the output device in the correct order.

This solution requires an enormous amount of memory. On a black-and-white plotter with 300 dots per inch, if we wish to make a plot 18 by 36 inches, we require 18 megabytes of storage. Computers with virtual memory can allocate blocks of storage that size, or we can work directly to the disk, but in either case the time to fill the raster will be large, dominated by disk access time. In a virtual memory system, the working set will be very large, so we will thrash. In a disk-based implementation, we will be continually moving our read/write location. It amounts to the same thing.

Buffering and Sorting

We can limit the size of the raster by building several smaller rasters in the order they are needed for the plotter. We generate the raster for a piece of the layout, send that buffer and start next one. The thrashing problems are solved, since we have much less data in memory at once.

The sorting we do is similar to the bin sorting we had for pen plotters. In this case our bins divide the cell into pieces that we can fit in memory and send the plotter when we are finished. Assume we want to plot minimum-X to maximum-X on the plotter. We sort the data by minimum-X values and maintain an *active* list containing all instances that are visible between the minimum and maximum X values for the current buffer. When we start a buffer, we add all features whose minimum X values are in the new buffer and remove those whose maximum X value is less than the minimum for this buffer. We generate the raster for the buffer, send it to the plotter, and move on to the next buffer.

We can choose the size of the buffer by the amount of memory we wish to devote to it. In fact, we can shrink the buffer to a single line, one X coordinate. When we do so, the plotting algorithm mimics the edge-based operations we will see in Chapter 8 for layout verification. Edge processing algorithms can be implemented efficiently and the same code that we use for layout verification can be used to convert data for rasterization.

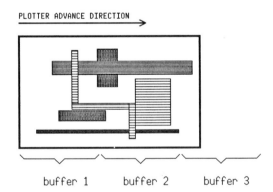

Exercises

Programming Problems

1. Implement a plotter for layout data as described in this chapter. Include commands to set the window and viewport and to select layers to be plotted.

2. Write a command scanner that takes as arguments an array of strings for the commands and a string for the current input, and returns an integer for the number of the command that matched. Unambiguous substrings should be legal. Use procedures from the layout parser if you wish.

3. Write a procedure to adjust the scale when zooming so every grid dot always appears exactly on a pixel.

4. Add a command to set the window scale by giving the scale at which the cell is to be plotted, for example 300x.

Questions

1. Under what conditions will a bounding box check slow down processing of the data?

2. How much memory do the list data structures require? How big is the largest chip you can plot? What happens when you run out of memory?

3. Describe two strategies for handling very large chips without running out of memory. What performance penalties do they incur?

4. Different companies have different colors for mask layers. How would you allow a user to modify the mapping of layers to colors.

5. How would you modify the plotter to allow a user to plot the cell on the graphics display then use the pointing device to specify other areas to plot?

6. When converting data to raster for a raster plotter, what is the optimum buffer size?

7. Natural clustering and buffered optimization have one significant advantage over the sorting methods for ordering data for vector plotting. What is it?

References

J.D. Foley and A. Van Dam, *Principles of Interactive Computer Graphics*, Addison-Wesley, 1982

M.D. Prince, *Interactive Graphics for Computer-Aided Design*, Addison-Wesley 1971.

CHAPTER 5

LAYOUT EDITOR

Although many tools have been written to simplify and automate integrated circuit layout, the layout editor is still the workhorse of integrated circuit synthesis. A layout editor allows a user to specify graphically the shapes that make up his chip. The concepts involved are simple and powerful -- a user draws what he wants to see on the chip. He needs only a set of design rules and a little experience with layout to use the tool.

The goal of this chapter is to introduce the concepts of a graphical editor as embodied in a layout editor, building on the software we've already seen. We use the plotter as a starting point, add graphical interaction and the ability to modify the data structure. Later sections describe alternative data structures, and different types of the graphical editor.

Overview of a Layout Editor

A layout editor is similar in many respects to screen-oriented text editors with which you are familiar. The purpose of the layout editor is similar, it allows a user to modify the data with intuitive and powerful operations. Thus, the user interface is a vitally important part of the design of a layout editor. In this application, a poor user interface is more devastating to users than poor algorithms. A poor algorithm causes delays, a poor user interface causes errors. A second important feature of the layout editor is the storage of the two-dimensional data in memory in a form that allows easy access for editing operations. We will investigate both issues in this chapter.

A layout editor consists of a command loop that accepts input from the keyboard and the pointing device, a module for operating on the data structure, a graphics module for display, and facilities for reading and writing. The command loop we use for the layout editor is virtually unchanged from the plotter's command loop that we saw in Chapter 4, though we will change it slightly to accept graphic input. The graphics module, parser and semantics are the same as those we used in the plotter.

Turning the Plotter Into an Editor

The plotter has nearly all the facilities we need to edit a layout file: it can read a file, building the data structure and can display the data on the screen. We add facilities to modify data and write the layout file when we are done.

Adding Data to the Data Structure

The task of adding data to the data structure is no more difficult than it was to add data when we read files in Chapter 2. The commands we define here to create data in the cell are analogous to the statements in the layout file.

The first step is to select a cell for editing. If the cell doesn't exist, we make a new record for it and make it the current cell, just as we did in the layout semantics module. If the user selects a cell that already exists, we make that cell the current cell. Similarly, when the user chooses a layer on which to create layout, we set the currentLayer to the new layer. If we are already editing a cell, we just stop editing that cell and start editing the new one. We don't need a separate end cell command.

Continuing to follow the layout data file, we can define commands to make boxes, polygons and instances:

```
addbox    <l> <b> <r> <t>
addpoly   <x1> <y1> <x2> <y2> ... <xn> <yn>
addInstance <cellName> <x> <y> <orientation> <instName>
```

When we see an addBox command, we make a new boxClass, fill it in and add it to the list of data in the cell, just as we did in the parser:

```
[editCommand] begin
  cellName := getString;
  currentCell := lookupCell(cellName);
  if currentCell=nullPointer then begin
```

```
      currentCell := new(cellClass)
      currentCell.name := cellName;
    end;
  end;

  [addBoxCommand] begin
    b := makeBox(getNumber,getNumber,getNumber,getNumber);
    putInList(currentCell.data,b);
    displayBox(b);
  end;
```

Deleting Data from the Cell

We can add commands to delete data from the cell, such as:

```
delBox <l> <b> <r> <t>
delPoly <x1> <y1> <x2> <y2> ... <xn> <yn>
delInstance <cellName> <x> <y> <orientation> <instName>
```

When we get the delete command, we scan through the list of data in the current cell and remove the box, polygon or instance that matches. If there is no match, we give a message and don't delete anything.

Of course, we do not need all this information to delete an object. We should be able to delete an object just by giving one piece of information such as its location. We only need one command to delete the object closest to a point:

```
  [deleteAtCommand] begin
    getPosition(x,y);
    b := getClosest(currentCell.data,x,y);
    if b <> nullPointer then begin
      removeFromList(currentCell.data,b);
      reDisplayArea(currentCell,b.bb);
    end;
  end;
```

To save display time, we only redisplay the area that changed instead of the whole cell. We would like to limit as much as possible the amount of data we redisplay, so if possible, we also limit the redisplay to the layer of the object we deleted. If we delete several objects, we merge their bounding boxes and redisplay the merged area.

Pointing As an Alternative to Typing Positions

Typing positions will immediately become tedious, and we have a much better way to indicate a position: we can point. The modification we make to `deleteAtCommand` is in the procedure `getPosition`. Instead of
   `x := getNumber;   y := getNumber;`
we invoke the procedure `userInput` in the graphics package to get the position of the cursor. The cursor position is in virtual screen coordinates, so we reverse transform the coordinates using the current display transform to get user coordinates:

```
userInput(pointing,ignore,sx,sy);
screenToUser(sx,sy,x,y);
```

This code gives us the location in user coordinates where the user pointed the cursor. We now use the point just like we used numbers that were typed. We can use this method to get locations for commands like the `box`. We can make similar modifications everywhere we need a point: in polygons, in zooming and so on.

Snapping the Points

Suppose we we are zoomed so one lambda is represented by ten pixels, and the user draws a box from 10,10 to 20,15. Suppose now he wants to extend the box on the right by drawing from 20,10 to 30,15. If his finger shakes just as he presses the button on the pointing device, he might miss the edge of the box by one pixel, .1λ. The box he gets runs from 20.1,10 to 30,15, leaving a small gap between the boxes.

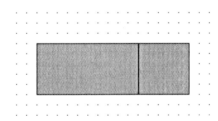

The cell may look good on the display, because he may be unable to see the one pixel difference, either because his eyes are not that good or because he is using stipple patterns anyway and can't see the edges of the boxes precisely. The problem is aggravated further because we are transforming coordinates from screen units to user coordinates using transformations that are susceptible to roundoff error. When rounding occurs, the two rectangles may appear to touch, but not touch in reality. Zooming-in would show the problem, but we cannot rely on users examining all their boxes in such detail.

To avoid this problem, we impose a grid in user coordinates and *snap* on the points to the grid (Sutherland 1965). That grid is usually much larger than the size of a pixel on the screen, but may not be, depending on the zoom factor. If we set the grid to 1λ, then the location 20.1 will be snapped to 20 anyway and the boxes will always touch.

Another problem with pointing is that people are not very good at pointing to the exact place they want. When you point near an edge of a box, you probably want to point at the box itself. It would be a poor system indeed that let you point one pixel off from a box and left a little gap there. That gap may be huge if you are zoomed out far. You would see the gap when you zoomed in. Instead, the system may *snap* the points to locations that are more meaningful. If you point within a few pixels of the edge of a box, the system should give you the coordinates of that edge, even if you missed it by more than the grid size.

The pointing grid and the maximum distance to snap to an object should be parameters that a user can change. That way, if they get in his way he can turn them off.

Filters

Our simple delete command can often be too ambiguous, since many objects may overlap at a point. We would like to use more criteria than proximity to select the object to be deleted. For example, we can only delete objects on a layer, we can limit the deletions to only those layers we are displaying on the screen or only objects wholly contained within the display area. These selection filters are applicable to any command that changes an existing object.

Editor State

In the last chapter, we discussed options for plotting selected layers of the cell, choosing grid size, and so on. These options, the filters for selecting objects, the default layer on which to create layout and others, make up the *editor state*, the set of default values that the editor uses so a user does not need to give all details about every object with every command. The editor state simplifies editing because the commands themselves are simpler, but the editor state is hidden from the user, becoming a potential cause for confusion. It is important that a user be able to view his editor state. Some of the state, such as the display options are easy to view; but others, such as the current layer, are not as obvious and are of such importance that they should be continuously displayed. Others pieces of state may require special commands to view them. The appropriate treatment of editor state is essential to provide an effective interface to a user.

Modes and Modeless Editors

Let us consider the actions a user performs to make and delete objects. To make a box:
- type B
- point to the lower left corner
- point to the upper right corner

After the user types B, the system is waiting for the cursor events to indicate the corners of the box. The state of the system is called a *mode*. Generally, the more modes your system has, the more difficult the system is to use because a user must remember which mode he is in. For example, if a user typed B by accident, he might not realize that the system is waiting for him to point to the corners of the box. If there is no prompt message on the screen, there may be no indication that the system is waiting for that kind of input. The user would try typing commands that would be ignored because the system is waiting for him to point. Eventually he may point to something else and get a box he did not expect. This confusion is very bad and is a major cause of error and frustration.

Modeless Editors

One solution to the problems caused by modes in an editor is to make a *modeless editor*. A modeless editor has no modes, a user can always give any command. Users may point anywhere at any time, and the system accepts the point. Commands take effect immediately and require no more inputs to complete their task. When a user gives the box command, the system finds the last two points and the current layer and uses them to make the box. If he wants to delete an object, he types D, the system uses the last point to determine what to delete. We trade modes in the editor for more state in the editor.

The difficulties with modeless editors are twofold. First, some commands may require a large amount of data to be pre-specified. For example, to make a polygon without modes, we must save however many points the user types. Modes allow us to give better feedback. If we know in advance that we are making a polygon, we can draw the outline of the polygon as the user indicates the points.

The second problem with modeless editors is that they require that all commands be one button press (or else you are back in a mode waiting for further keystrokes). This problem is not too important considering the few commands we have already considered, but we quickly run out of mnemonic characters to assign to a command. For example, if you zoom in with "Z" , then what letter do you use for zooming out? The problems with modeless editors increase as we build a more graphical interface.

Using the Pointing Device to Invoke the Command

We get a position from the pointing device when we press one of its buttons. If we assign functions to buttons on the pointing device, we can simultaneously give the command and indicate a point for the command (Fairbairn and Rowson, 1978). For example, if a button on the pointing device meant `delete`, then we can point to an object and press that button, deleting it. Unfortunately, we have many more commands than buttons on the pointing device.

If we want to maintain a modeless editor, we have two options: define shift keys on the buttons or redefine the functions of the buttons. A shift key for the pointing device works just like the shift key for a keyboard: when the key is depressed, the buttons on the pointing device have a different function. We decode the function by examining the button pressed and the key from the keyboard.

On most keyboards, though, we can't tell if a key is being held down, but we can provide keyboard commands that change the meanings of the buttons on the pointing device. So the buttons on the pointing device may mean: `deleteAt`, `zoom in` and `zoom out` originally, and change to `addBox`, `addPolygon` and `addConnector` after we hit the letter A (for Add). Of course, we still encounter some difficulties with coming up with mnemonic "shift" letters.

Most editors settle for a combination of features of modes and state, and are forced to do so if the pointing device has only one button. In this case, the user selects the meaning of the button first by typing or by selecting the command from a menu on the screen. He invokes the function by pressing the button on the pointing device. The editor has a *current command mode* as part of its state.

Modes are not inherently bad: any program you run is, basically, a mode. On nearly every computer, we move from the system shell mode to a text editor mode. We switch from a layout editing mode to a plotting mode to some other mode as we run different programs. In general modes are acceptable if:
1. there is always some indication on the screen what the current mode is,
2. modes are not nested too deeply: there are no modes within modes,
3. you exit all modes the same way, so users can leave a mode without knowing what mode they are in, and
4. you can get to any other mode and back quickly.

Menus As an Alternative to Typing Commands

A common practice is to put the list of commands on the screen in a menu, dividing the screen into an editing area and a menu area. Pointing to a menu entry is equivalent to typing the command. Thus, we can eliminate most of the typing that a user would have to do. He need never take his eyes off the screen or his hands off the pointing device.

The problem with putting the menu on the screen is that it takes screen area, leaving less space for display and editing the cell. A solution to this problem is a popup menu, described in Chapter 3. When the user presses one of the buttons on the pointing device, we bring up a popup menu under the cursor. When he selects an entry, we take away the popup menu. We will return to the tradeoffs between different methods of giving commands later in this chapter when we examine user interfaces.

Data Structures

The data structure we have discussed is a simple one: a list of rectangles, polygons, instances and connectors. It is simple and versatile. Additions are fast, but deletions take time proportional to the amount of data in the cell because we must search the data for the object to be deleted. In this section, we describe different kinds of data structures that have been used in graphics systems and discuss their advantages and disadvantages. We do not develop this comparison to one clear winner, the data structure that works best depends on the amount of data in a cell and the kinds of operations performed on that data.

Searching and refreshing the list-based data structure can be slow. Consider that we wish to find what, if anything, is at a certain location.

We step through the data one object at a time, checking bounding boxes. If the box surrounds the point, then we add that object to the list of objects at that location. We traverse the whole list of objects in the cell. That list can be very long.

Instances make searching and refreshing more complex but can speed those operations considerably because instances provide grouping for objects. Normal searches in a cell stop at the instance. When displaying, we look inside instances, but we can eliminate from consideration all objects in instances whose bounding boxes lie entirely outside the display area.

If we wish to search all the way to primitive objects, we look inside an instance by transforming the point back to the coordinate system of the defining cell and search that cell's list. We apply the process recursively until we finish.

```
procedure whatIsAt(pointer(cellClass) cell;
  real x,y;  modifies pointer(list) found);
begin
  real cx,cy;
  for all elements: e in cell.objectsList do begin
    if e.boundingBox.contains(x,y) then begin
      case e.type of begin
        [boxType] [connectorType] putInList(found,e);
        [polygonType]   if polygonContainsPoint(e,x,y) then
                        putInList(found,e);
        [instanceType] begin
          inverseTransformPoint(e.trans,e.dx,e.dy,x,y,
            cx,cy);
          whatIsAt(e.cell,cx,cy,found);
        end;
      end;   # case
    end;
  end;
end;
```

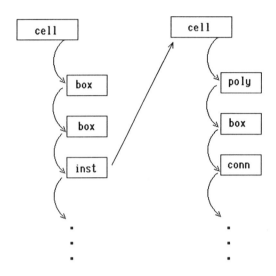

Hierarchy does not save us from having to visit every object in the cells data list, though. We can reduce that time if we sort the data, but the choice of search key is not obvious. Should we sort by x-value or y-value? By increasing or decreasing coordinates? By top edge or by bottom edge of the bounding boxes? The discussion that follows gives data structures that have been successful in graphical systems. The questions we want to ask while traversing are:

- "Which database element is the first one to use this coordinate?"
- "Is there no hope of ever finding this coordinate again in the remainder of the database?"

Assuming we are sorting in decreasing Y, these questions become:

- "Which is the first object that has this Y value at or above its bottom?"
- "Which is the last object that has this Y value at or below its top?"

If we sort by bottom Y then we can do a quick $\log(n)$ search to answer the first question, but we must search to the end of the list to answer the second. On the other hand, if we sort by top Y, then we cannot know which object in the list is the first to be displayed, but we do know when we can stop scanning the list. There are many options, including sorting by multiple keys and keeping a more complex data structure. We will now investigate data structures that attempt to address the weakness of simple sorting.

Unions

Let us return to our discussion of hierarchy. We found that we could use the bounding boxes on instances to limit our search. The bounding box and hierarchy provide some two-dimensional structure on the data that allows us to optimize our algorithms somewhat. There is really no reason why we need instances to implement this kind of structure: we can collect nearby objects together into *unions*. A union is analogous to a compound statement in a programming language and can be thought of as a cell without the hierarchy: just a collection of data with a bounding box. Conversely, a cell can be viewed as having two parts, the cell interface and one union containing the cell's data and bounding box. Users never see unions, we add them to our data structure to improve the performance of our algorithms. A union allows us to ignore collections of data by checking their cumulative bounding box. A data structure with unions looks like this:

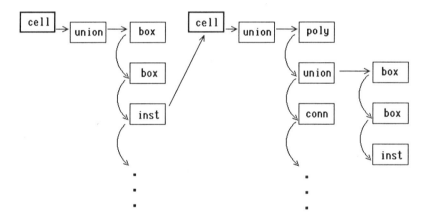

When searching for the objects at a point and we find a union, we check the bounding box of the union and, if necessary, search the list of objects in the union. To get the greatest advantage from unions, we build unions from objects that are close to one another.

If we force all objects in a cell to be part of a union and keep unions from overlapping, we have a bin-sorted arrangement, where the bins are the unions. We can choose the number of bins based on the number of objects in the cell or we have a fixed number of bins and recursively divide unions into smaller unions if it has too many objects in it. The result is that we visit $O(\log n)$ unions to find an object.

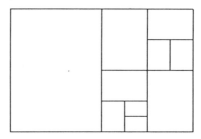

This leads to our next question: "when do we stop dividing data into unions"? If we divide unions until there is only one object in each one, the large number of bounding box comparisons would probably slow the performance of the system and require more comparisons than they saved. In addition, each union takes some space, enlarging the stored data for the cell leading to more disk I/O from virtual memory use. The correct answer to this question depends on the relative speed of the bounding box comparison and on the number of objects.

Two Dimensional Binary Search Tree

Consider the problems we find when making unions as above when we encounter a large object, such as a long wire. It falls in the regions for many unions. If we put it in a union, that union's bounding box will be large and it will have to be searched frequently. In this section and the following section, we develop tree-based data structures using the kind of locality we discussed with a hierarchy of unions.

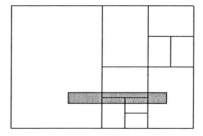

We can use an application of k-d trees developed by Bentley (1975) to address this problem. We start with a rectangle that includes all the objects to be displayed. We choose a point inside the rectangle make a vertical line through that point, splitting the cell into two regions. Objects wholly to the left of the point go into the left region, objects wholly to the right go into the right region. Objects that cross the vertical line are kept in the current region sorted along the cut line. We subsequently split both of the sub-regions with horizontal lines and continue splitting those sub-sub-region with vertical lines and so on. We stop splitting when there is no data in a region.

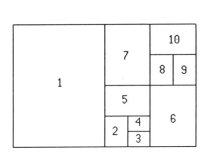 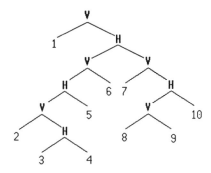

The resulting data structure is a binary tree. Each interior node in the tree represents either a vertical or horizontal split, and as we search down from the root, we successively limit our search area in the horizontal and vertical directions.

```
procedure whatIsAt(pointer(treeNode) n;
   real x,y;  modifies pointer(list) found);
begin
   if n.vertical then begin
     if x<n.splitLoc then whatIsAt(n.lowSon,x,y,found)
     else whatIsAt(n.highSon,x,y,found);
     searchListSortedInY(n.objects,x,y,found);
   end else begin   # n is a horizontal split
     if y<n.splitLoc then whatIsAt(n.lowSon,x,y,found)
     else whatIsAt(n.highSon,x,y,found);
     searchListSortedInX(n.objects,x,y,found);
   end;
end;
```

Quad Trees

A popular version of the two-dimensional search tree is the *quad tree* (Kedem 1982). A quad tree is formed in a similar way to the binary search trees we have just seen. Start with one area containing all the data in the cell. Divide the area of the cell into four equal-sized regions by splitting both the edges of the cell in half. Each of these regions are themselves split into four more regions. Each region contains a list of the objects that are contained in the region but not wholly contained in one of its sub-regions. The resulting data structure is an order-4 tree, as shown in the following figure.

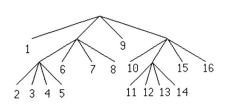

Now, when we search, we only need investigate the regions in the path from the root to the areas that are actually in the displayed area. Searching for the nearest item to a point is also efficient. This operation takes $O(\log n)$ time.

As a further improvement, we make two lists of objects: one that crosses the X dividing line, one that crosses the Y dividing line. An object may be in both lists. We sort the objects along those lines to further reduce the search time. Finally, we make a lower limit on the size of a region. Rather than make tiny regions, we make one list of all the objects in the region and search the list for the object we want.

```
procedure whatIsAt(pointer(quadTreeNode) n;
   real x,y;  modifies pointer(list) found);
begin
   if x<n.x and y<n.y then whatIsAt(n.llSon,x,y,found)
   else if x>n.x and y<n.y then whatIsAt(n.lrSon,x,y,found)
   else if x<n.x and y>n.y then whatIsAt(n.ulSon,x,y,found)
   else if x>n.x and y>n.y then whatIsAt(n.urSon,x,y,found);
   searchListSortedInX(n.horzontalList,x,y,found);
   searchListSortedInY(n.verticalList,x,y,found);
end;
```

The quad tree has fewer internal nodes than a binary search tree, so there is less storage overhead and fewer comparisons to reach the data at the leaves of the tree. Each node is handled identically, there is no separate treatment of the vertical and horizontal division.

The performance of complex data structures depends on the kinds of operations we perform on them. Although search time is improved, addition and deletion in tree-like data structures take $O(\log n)$ time instead of $O(1)$ time as in the simple list case. Further, to maintain good performance, we must keep the tree balanced. Rebalancing a tree can be an expensive operation, but integrated circuits are usually well behaved and balance is not usually a problem.

Large, complex data structures may degrade the performance of the system. Complex data structures perform better with large amounts of data, but simpler structures have less overhead and perform better with small amounts of data. The tree-like data structures require more space, since we must store the interior nodes in the tree in addition to the data. Larger data structures force more use of virtual memory. Paging due to virtual memory slows down the system.

Finally, maintenance of these complex data structures requires more time. The amount of complexity you can tolerate in your data structure depends on the relative speeds of your processor, disk and display as well as the size of your machine's physical memory and the expected amount of data to be shown at once. In practice, most systems adopt some sort of sorting, dividing the data into separate lists by layer and sorting in one dimension. Quad trees are used in applications that expect very large amounts of data.

Corner Stitched Tiles

Corner stitched tiles (Ousterhout 1984) is the name given to a totally different data structure oriented toward operations on areas in a plane. Corner stitched tiles represent the area of the plane itself, rather than objects in that area.

The area of the cell is divided into variable-sized *tiles*. Tiles represent the presence of a layer or open space. All the area of the plane is represented. Each tile has four pointers: pointing to the adjacent tile immediately below the lower left corner (b), immediately left of the lower left corner (l), immediately above the upper right corner (t) and immediately right of the upper right corner (r). These pointers allow algorithms to traverse the data structure, but in a data dependent way, since the tiles represent the data in the cell. To improve the efficiency of the traversal algorithms, all tiles are merged and split to form maximally-horizontal strips.

It is not necessary to store all four coordinates of a tile. Each tile can keep only its bottom and left coordinates because x.top=x.t.bottom and x.right=x.r.left. Of course, this cannot be done with the rightmost and topmost tiles, which represent the infinite space to the right and top. They are handled as special cases.

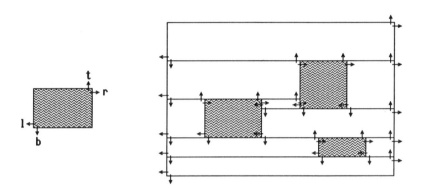

To find the tile at a given x,y location, we start at any tile and follow t and b pointers to find a tile whose vertical range contains the y dimension of the point. Then follow l and r links to find a tile whose horizontal range contains the x dimension of the point. Because the tiles are not uniform size, horizontal movement may have misaligned the vertical position. Therefore, we iterate between the two pointer-following steps until we find a tile that contains the point. In the worst case, this requires visiting $O(n)$ tiles, where n is the total number of tiles in the system, but the expected performance is only $O(\sqrt{n})$ tiles. Since we must search for a tile when inserting and deleting, all operations have this same time complexity.

```
# find the tile at x,y
procedure whatIsAt(pointer(tile) startTile;
   real x,y;   modifies pointer(tile) found);
begin
   found := startTile;
   do begin
     if tileContainsPoint(found,x,y) then return;
     do begin   # vertical scan
       if found.bottom>y then found := found.b
       else if found.top<y then found := found.t
       else done;
     end;
     do begin   # horizontal scan
       if found.left>x then found := found.l
       else if found.right<x then found := found.r
       else done;
     end;
   end;
end;
```

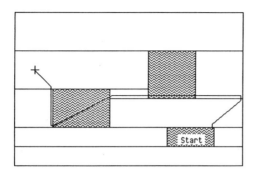

The average behavior of this algorithm is significantly improved if the starting tile is near the destination. We can make a good guess at such a tile in a layout editor if we start at any tile on the screen, since there is a good chance that a user is pointing to a tile on the screen.

The discussion has covered one layer with white space. A cell could be represented as one tile layer for each mask layer. Layers that do not intersect may be represented in the same plane. For example, the Magic layout editor (Ousterhout et al. 1984) represents polysilicon and diffusion in the same plane. Their intersections are stored as separate tiles with a tile type indicating a transistor. This dynamic recognition of transistors in the data gives the system many of the advantages of *symbolic layout*, which is discussed later in this chapter.

The corner stitched data structure shows its power when doing adjacency checks such as polygon merge, as discussed in Chapter 4. Corner stitched boxes have nearest-neighbor connections that are required operations for design rule checking and circuit extraction, discussed in Chapter 8. The algorithms for scanning corner-stitched tiles to check the separation between tiles and for accumulating all the tiles on a given layer are simple iterative algorithms.

Essential Features

The goal of a graphical editor is to capture the design form the design. The commands in the editor make up the language the user has to express the design. In this section, we focus on features and commands in the layout editor. In the following section, we address the problem of tying together the commands into a coherent user interface.

Creating Layout

Set Layer. As we already discussed, we can set the layer on which users create new layout. Many systems allow a user to create layout on multiple layers simultaneously. This allows a user to draw, for example, the multiple layers of a contact simultaneously.

Draw Box. Many successful systems have limited their users to this one primitive, and thus to *orthogonal layout* consisting of only horizontal and vertical edges. This restriction does not simplify the layout editor significantly, but it does simplify design rule checking and circuit extraction algorithms.

Make Connector. Connectors are not essential for drawing layout, but they are helpful for connecting cells without viewing all interior layout. They set names of nodes for simulation and are indispensable for automatic routing tools.

Make Instance. Making an instance has three parts: specifying the cell of which to make an instance, adding the instance to the database, and drawing it on the display. A user can specify an instance by typing the name of its defining cell or by pointing to the name in a menu of cell names. Many systems allow a user to specify the orientation of the instance when it is created, either directly or by having a default orientation that a user may change (for a modeless editor). A third possibility is to always make instances in the standard orientation, requiring users to change the transform after creating the instance.

Make Array. An array specification creates many instances laid out in a rectangular array. The command takes the number of times the instance is replicated in X and Y and the spacing in X and Y. Commonly, alternate instances may be mirrored about the vertical axis in increasing X, and about the horizontal axis in increasing Y. The array command may generate instances or, for efficiency reasons, the array may be stored internally as a single object.

View Hierarchy Level. Display of all the contents of all cells can be time consuming. We provide a display option to show an instance as a blank rectangle, called a *phantom* or *icon*, to save time displaying its contents. We can optionally display the cell's connectors and connector names with the blank *phantom*. Since users cannot edit the contents of the defining cell of the instance by pointing at the instance layout anyway, an opaque rectangle is a reasonable way to view the instance. Users must still have the option of viewing all the contents of all cells and they should have the option to select the number of levels of hierarchy that they view.

Editing Layout

Move. Move an existing piece of layout to a new location on the screen.

Stretch. Stretch moves the edge of a box, keeping other edges where they are. This allows a user to change the size of a box, making it bigger or smaller as needed.

Delete. Remove objects from the cell.

Transform. Change the transform of an instance.

Change. As a general rule, it should be possible for a user to change anything that he could set with the data creation commands. Users should not be forced to delete and re-enter layout just to change a single attribute. A separate command to change features of object would serve to change a connector's name, width and layer; an instance's name and transform; and an array name, transform, replication counts, offsets and mirroring options.

```
☒◻                      Array
 accept
 x reps          5
 x goes          right left
 x spacing       22
 y reps          2
 y goes          up down
 y spacing       42
 display options all edge corner none
```

Viewing Options

Zoom in/out/cell. Change the viewing scale and position. Zooming *in* gives a more detailed view of a smaller area. Zooming *out* gives a less detailed view of a larger area. Zooming to *cell* adjusts the viewing parameters so a user can see the whole cell at once without knowing the size of the cell in advance.

Snap Grid. As we already discussed, it is advantageous to have a grid on the coordinates to which all pointing operations snap. Users must be able to alter the grid spacing.

Display Grid. We use a second grid as an editing aid. This grid is visible on the screen to allow a user to see dimensions and distances to facilitate correct sizing and alignment of layout. Users must be able to change the grid spacing as well as turn the display grid on and off.

View Layer on/off. Modern fabrication processes can have dozens of layers. We will add additional layers for our own use in the tools. The result is that all designs, when viewed in their totality, consist of a confusing mass of shapes, lines and edges. It is frequently necessary to turn off viewing many of the layers simplify the view of the layout and ease the user's job of interpretation.

Reading and Writing

Most of these commands are analogous to the commands in text editors.

Save Edits. Save changes in a layout file.

Load a File. Load cells from a layout file.

Cancel Edits. If a user decides that the changes he made are bad he may discard them. We restore the cell by re-loading it from the last saved version.

Edit Cell. Edit a different cell in the same file. We can allow a user to indicate this cell by pointing to an instance of it, by typing its name or by selecting its name from a menu of cell names.

Quit. The system must not let uses accidentally leave without saving changes. If cells have been changed, but not saved, the system should ask the user if the changes should be discarded and give him the option of remaining in the system to save his changes.

Desirable Features

These features are not essential for the layout editor to be useful in producing chips, but they are often very important in facilitating the design process.

Ruler. We cannot rely on the grid to always be exactly the size and position for a user to judge distances. A ruler lets a user point to two positions and types the coordinates and the distance between the two points. This feature may also take the form of a dynamic display mark with tick marks for distances.

Flatten. We deal with a hierarchy, but often that hierarchy must be rearranged. This command removes an instance, replacing it with its contents, transformed to the same location and orientation in the current cell. The result is the removal of one level of hierarchy with no change to the layout of the cell.

Make Cell. This command is the opposite of flatten. The area of the cell indicated by the user is collected into a cell of its own and replaced with an instance of that cell. This creates one level of hierarchy.

Rectagon. Rectangles are easy to deal with as a tool builder. However, designers do not always think of their shapes as rectangles. A *rectagon* is a polygon that has edges parallel to the coordinate axes. A rectagon can always be decomposed into a set of rectangles, but this command provides a more convenient way to draw rectagons.

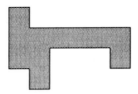

Non-orthogonal Angles. Rectangles and rectagons are simple geometric features to model and edit. But many designs can be made much more dense by adding lines at non-orthogonal angles. In digital designs, 45-degree lines are usually sufficient, so systems that allow angled lines frequently limit them to 45-degree lines so they can be handled efficiently as special cases. A separate command can allow any-angle edges or restrict polygon angles to be multiples of 45 degrees or 90 degrees (rectagons).

Cut and Paste. Many operations require users to move, copy or delete whole areas of layout. The *cut* operation removes an area of layout, placing it in an internal buffer for temporary storage. A *copy* command copies the area into the buffer without removing it from the cell. A later *paste* copies the data from the internal buffer and places it in the layout at the cursor location. Cut and paste allow a user to deal with the layout in terms of areas instead of objects. The areas are usually more intuitive, since conceptually, the user is manipulating a picture of areas in the layout, regardless of the set of rectangles and polygons he uses to express it.

Edit In Place. To get the cell interface right, users must frequently edit a cell while viewing the environment of an instance. One way to accomplish this is to copy a section of the environment into the cell, edit the cell and cut the environment layout. A much cleaner way to do it is to allow users to select an instance to edit, and to edit the defining cell of that instance while pointing at the instance. To do this, we reverse transform the user coordinates with the instance's transform to get positions in the cell's coordinate space. Users may experience some confusion between the cell and its environment, so we may display the environment with a duller color scheme, perhaps with a less dense stipple.

Wire. A major part of integrated circuit layout is concerned with wiring together already-designed cells or transistors. The wire command lets a user specify layout by drawing a center path, while the system fills in rectangles (or polygons) with a user-settable width along that path. A wire of width 3 from 2,2 to 6,2 looks like the subsequent figure. When a user turns the corner and continues the wire, as shown on the right, we continue the wire and fill in the outside corner.

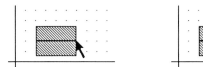

Node Name. Users may wish to name nodes in the circuit for documentation or simulation purposes. We can extract node names from connectors, but there may be important nodes that do not correspond to connectors.

Undo. When a user makes the wrong editing operation, he requires a single command that removes the last change. Otherwise, he must carefully remove layout he accidentally created, add layout he accidentally deleted and undo any changes he made to connectors and instances. Undo enhances the usability of a system considerably. To implement the Undo, we keep lists internally of added, deleted and changed objects and replace them on command.

User Interface

A *user interface* is the name we give the whole method of interaction between the system and the user. The interface has two parts: the way commands are specified and output is shown to the user; and the actions of commands, the manner in which they create data. The division is roughly the same as the division between syntax and semantics of a language. In this case, the language is the language used to communicate with the design tool.

In the beginning of this chapter, we developed a user interface bit by bit, including graphical editing features with our typed commands. This incremental method is perhaps the worst way to develop a user interface.

It is very important that the interface have some underlying concepts that simplify its use. Do not underestimate the importance of the user interface. After all, our task is computer *aided* design. The user interface must be designed as a coherent unit.

Typically, a user interface has one or two fundamental underlying concepts. These concepts provide the framework for organizing commands and dictate the way users view and manipulate data. Newman and Sproull (1979) discuss the overall considerations to avoid bad user interface design, but a "good" interface is somewhat a matter of personal taste. This section describes the interfaces of layout editors from two different institutions. The developers of the two editors faced different constraints imposed by their expectations about their users and the users' goals. We describe the user interface of the tools and the underlying reasons for the decisions that drove the user interface design.

Caesar

Caesar (Ousterhout 1981 and 1984) is a layout editor developed at University of California at Berkeley. The developers of Caesar believed that screen area was most efficiently used by devoting all of it to display area. So Caesar has no menus, the commands are all short sequences of characters typed on the keyboard. Single-letter commands take effect immediately, longer command words are preceded by a colon and terminated with *eol*. Although novices find the abbreviated commands confusing, experts can edit layout very quickly.

Caesar only deals with boxes in its data structure, so users are restricted to orthogonal layout. The advantage Caesar gains by this restriction is a very uniform, simple interface.

The editor we described at the start of the chapter works on objects, pieces of layout like boxes and polygons. Users create them, move them, alter them and delete them. In Caesar, instead of indicating the objects in to be deleted, a user indicates the area to be cleared. Instead of creating boxes, Caesar fill areas with various layers. This is called a *painting* metaphor, since a user paints areas with layers, rather than covering those areas with objects.

Caesar uses the pointing device to set a rectangular area on the screen called a *cursor*. Users create a box by setting the rectangular cursor and pointing to an area of the screen that already has the layers that they want in the cursor area. Caesar fills the cursor area with those layers.

Users use *cut, copy* and *paste* to delete, move and copy layout. On a *cut*, Caesar removes from the cursor area all layout on the selected layers, regardless of the number of boxes inside the area. This cut may require that Caesar split rectangles in its database, because it only removes the layout under the rectangular cursor. After *cutting* layout from an area, a user can move the cursor and *paste* the area back in another place. He can use cut and paste to move an area or copy it: simply paste it twice. Caesar lets a user restrict the cutting operation to specific layers for selective operations.

Because users deal with areas and not objects, Caesar optimizes the creation of boxes in its data structure. When a user makes a box, Caesar merges the new box into its list of boxes. Caesar does not allow two boxes to overlap, thereby speeding up its cutting and pasting operations. Although the data structure remains consistent, the addition of non-rectangular features, such as polygons and point-like connectors, must be handled separately. An obvious technique is to make a separate list of non-rectangular items, separating connectors, instances and polygons from rectangles. A separate search is necessary for these lists, but they will usually be short.

This separation of the internal storage of the data from the user's view of the data is an important one. It allows the data structure to be optimized internally without changing the way the user deals with it. Investigation of a good data structure for Caesar led to the development of corner stitched boxes in Magic (Ousterhout et al. 1984), the successor to Caesar.

VTIlayout

We see a much more complex interface in VTIlayout, a layout editor marketed by VLSI Technology, Inc. (1986). It is part of a larger collection of tools that share the same interface. Each tool resides in a *window*, a rectangular area of the screen (a viewport). The windows may be drawn any size that is convenient for the user, though they are usually drawn to fill the entire screen. Windows may overlap, but the active window, which receives all commands, is always fully visible. If the user points to a window that is not active, that window becomes active and is redrawn on top of all other windows.

VTIlayout has many commands and options, leading to a large number of keywords for users to memorize. The developers of VTIlayout wanted to minimize the amount of typing done by a user, so all commands are invoked by pressing buttons on the pointing device. VTIlayout makes extensive use of menus, and the keyboard is used only for setting names. Typing is minimized in favor of pointing.

Layout Editor

The three buttons on the pointing device are called *popup*, *mark* and *doit*. The *popup* button brings up a popup menu of commands under the cursor. When you select a command from the menu, the menu disappears and the command you selected is your *current* command. When you press the *doit* button, VTIlayout executes the current command. The *mark* button sets the *mark*, the equivalent of the cursor in Caesar. In VTIlayout, the mark is either a point or a rectangle, depending on the current command. If the command requires one point, it uses the location of the cursor. If it requires two points, it also uses the mark. If it requires a rectangle, it uses the rectangular mark.

Infrequently-used commands and commands that execute without changing the current editing mode (i.e. pan, zoom) are invoked by pointing to the fixed menu on the right side of the window. Those menu entries may invoke commands or bring up popup menus of their own.

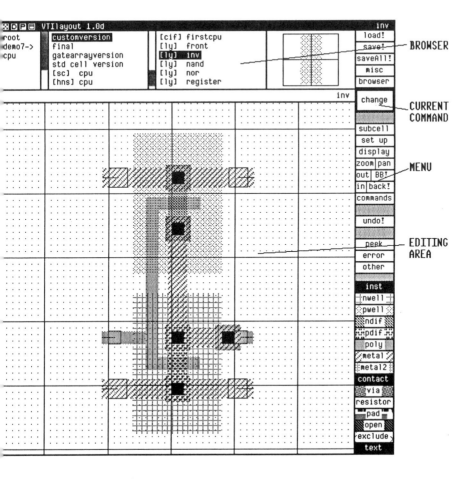

Even cell names need not be typed. At the top of the window is a *browser*, a hierarchical list of cells organized by the user into libraries and categories. The organization of the browser is similar to a hierarchical file system with libraries and categories as directories and cells as files. The leftmost pane in the browser shows the path through the hierarchy to the current category being viewed. The next pane shows the contents of the category that contains the current category and the third pane shows the contents of the current category. A user may point to a new category or library to make that one current. To load a new cell for editing, a user points to the cell name, then invokes the load command by pointing to that menu entry on the right.

Not all typing is eliminated, though. Users still type cell names when setting the name of a cell or a connector. Typing is minimized, but not eliminated, since text entry is necessarily a typing problem.

The VTIlayout data structure includes boxes, polygons, connectors, instances and node name points. VTIlayout supports a painting interface, much like Caesar, and merges new rectangles and polygons into its data structure as they are drawn. Cut and paste operations like Caesar's perform the basic graphics manipulation task.

VTIlayout does not restrict users to orthogonal layout. Users may draw polygons and select orthogonal, 45-degree limits or no limits on the angles. Internally, polygons are translated into boxes if possible. Arbitrary-angle polygons are stored in a separate list from the boxes, so there is no performance degradation for users who do not use them. When a user cuts or pastes an area, VTIlayout modifies objects in both lists.

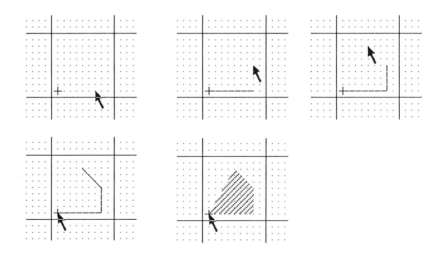

VTIlayout allows users to draw wires as an alternate way to add boxes to the data structure. The user specifies a path and the system makes boxes of a predefined width to cover the path users specify. Wires are not retained in the data structure as wires and once made, they are boxes just like other objects in the layout and can be manipulated with the same cut and paste operations.

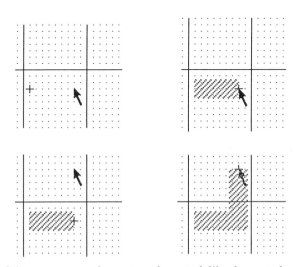

Connectors and instances can be cut and pasted like layout, but they have properties beyond the simple layout properties. Connectors have node names, instances have a defining cell, orientation and so on. To set these attributes, a user uses object-based operations. Using the *change* command, a user points to an instance or connector to bring up a dynamic display *property sheet*. The property sheet has pairs of attribute name and value. A user points to the attribute he wants to change and types a new value or points to new value in a list.

The Database Alternative To Reading and Writing File

When editing as we described, we make a copy of the cell in memory, make the changes, then decide whether or not to keep those changes. If we keep the changes, we modify the original in a single operation. This model of editing has some disadvantages. First, if the computer crashes before a user saves his work, he loses all his changes since the last time he saved it. In addition, if there is a project where many designers wish to modify the same cell at the same time, we have to implement some sort of check-in and check-out system to prevent one user from accidentally wiping out changes made by another.

These problems have been already addressed by transaction-oriented shared databases. In such a system, all cells and all objects in cells are stored in a single, shared database. When a user edits a cell, the change is immediately recorded in the database and transmitted to all users who are using the database. Many users can simultaneously access the database. The data is kept consistent because all users' views of the data are constantly updated from the database. The database system may also keep transaction records, so any number of changes can be unravelled to revert to a prior layout.

These database operations replace the reading and writing operations we described earlier. However, they are not without cost. The code of the database can easily exceed the size of the code for our application. This is not a problem if we use a commercially-available database package. However, this solution requires that we fit our data into the database form. Proper selection of the mapping of our data into the database is essential for good performance. Regardless, the performance of database systems will always be inferior to simply changing fields in records in memory, and our system will have to deal with the data in form for the database system, which will be larger than it would if there were no database. Although many systems have been written using databases, current computers are not powerful enough to offset the performance degradation one finds with such an interface.

Symbolic Layout

A layout editor is a general tool, but it provides no checking that the picture a user has drawn is actually an integrated circuit. It is easy to leave off one of the layers on a contact or transistor, resulting in an unconnected or incomplete circuit. We can avoid some of these potential errors by predefining contacts and transistors and allowing users to place them as single symbols. We refer to transistors and contacts as *components*.

Suppose we restrict users to input only wires on conducting wiring layers and to place instances for components. Dealing with these primitive electrical objects as atomic symbols is the basis of *symbolic layout*. Because we can identify transistors, contacts and connections between them, symbolic layout allows us to recognize circuit features in the layout easily. We can build an internal data structure that includes a transistor-level netlist for simulation and optimization.

We can implement a simple symbolic layout editor using the layout editor we already have as a base, drawing on the object-based data structure. We define cells for the components and place instances for them. We eliminate the box and polygon commands and use only wires to connect the symbols. However, greater advantages can be had by providing a system more directly targetted to symbolic layout.

Viewing

We can improve the way we view a symbolic cell by defining transistor and contact *icons*, the symbols that appear when we do not view the contents of the cells. Instead of the blank box we proposed for the layout editor, we can show the traditional electrical symbol for a transistor or a schematized view of the layout of a transistor. We apply the symbolic view to wires as well, showing them as thin lines connecting the symbols.

Since the wires and components still have physical attributes of position and size, and since their layout is fixed, we can choose to view the cell as full layout, or as the symbolic view. This symbolic view of the circuit shows the transistors, contacts and wires, giving the user an uncluttered view of the circuit.

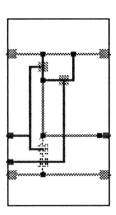 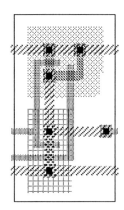

Wiring

Symbolic layout systems offer more advantages than a convenient method of specification. In order to effectively enforce a symbolic style of layout, we include checks on the wire commands to ensure that the wires a user creates do not accidentally overlap to make transistors. If a user accidentally specifies an illegal overlap, we flag it as an error and do not generate the wire.

We can also force wires to connect to components at their connectors. When a user ends a wire near a connector on an instance, we snap the point to the instance connector, make the wire and record the electrical connection in the data structure. We can easily generate a transistor netlist from the symbolic layout because all the transistors and contacts are explicitly called out as symbols and all connections are complete wires. Since the user explicitly connects to components, we can maintain those connections when the user moves contacts and transistors. We move the wire segments that are connected to the components to maintain the connection.

Symbolic Layout Compaction

If we know the minimum spacing rules for the layers in the fabrication process, we can write a procedure to move all wires and components to minimum design rule spacing, keeping all components connected. This process is called *compaction*, and systems that provide it are often called *sticks* systems (Williams 1977). There are two major kinds of algorithms used for compaction: *virtual grid* (Weste 1981) and *constraint graph* (Hsueh 1979) algorithms. Both methods restrict the layout to orthogonal features and space the cell one dimension at a time. The same algorithm is applied both vertically and horizontally. In both kinds of systems, a *compactor*, sometimes called a *spacer*, moves left to right, positioning each component and wire as far to the left as possible without causing a design rule violation, then does the same vertically, moving components down as far as possible.

A virtual grid system starts by examining the coordinates at which components and wires have been placed. Starting with the leftmost features, all features at the same X coordinate are moved together as far to the left as possible without causing a design rule violation from any one of them to the layout already placed. When the compactor is finished horizontally, it repeats the operation vertically, moving all layout down as far as possible. The result has no spacing violations.

This process assigns a new coordinate for each coordinate used in the layout. Since all features that use a coordinate move together, the designer must offset features that must move separately. The difficulty in virtual grid compaction is to properly offset features that should be moved independently to gain a dense compaction.

A constraint graph system builds a weighted directed graph in which the nodes in the graph are components in the symbolic layout cell and the arcs are the spacings between the components. Each arc has a weight that corresponds to the size of the spacing rule between the two features. The compactor assigns each node a position that is the length of the longest path into that node. The algorithm finds the *critical path* from the left edge to each node and assigns the features at that node to that position.

The performance of a constraint-graph compaction depends on the number of constraints in the graph. The important algorithms for constraint graph compaction are the ones that generate the graph, eliminating unnecessary constraints.

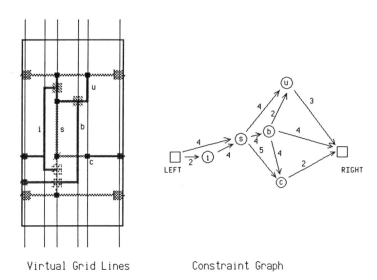

Virtual Grid Lines   Constraint Graph

This description of compaction is necessarily terse and does not address the many complications to the algorithms. For example, the compaction methods we described move all features as far left as they will go. This may unnecessarily stretch wires that would be better left short along the right side, such as the contact on the output line in the example. Also, fixed-sized features and constraints that users supply on the final positions may make a solution impossible, so the compaction may fail. Kingsley (1984) summarized the difficulties of graceful failure. For detailed description of symbolic layout and compaction, see Hsueh (1979), Cho (1985) and Weste (1981).

Additions for Symbolic Layout

Many symbolic layout systems have automatic transistor recognition. Instead of disallowing crossing layers that form transistors, they create a transistor at the intersection and connect the wires to it. Further, when a user draws a wire of one layer so it ends on a wire of another layer, the system automatically inserts the contact.

Since transistors are explicitly called out in the data structure, it is reasonable that users modify them as single features. A common operation is changing width or length of a transistor or the width of a wire.

Typically, symbolic layout systems do not store components as instances as we described initially. They keep them internally as primitive objects along with wires. The locations of connectors on components and jog points in wires are implemented as pointers to coordinate records. When a user moves a wire or a component, the system updates the coordinate records which are referenced by all components and wires, so wire lengths change and wires remain connected.

Symbolic layout systems provide the ability to view the layout of a cell as well as the symbolic "sticks" of the cell. They also allow a user to view the critical spacing path, the path through the features that constrain the size of the cell in both dimensions. This critical path indicates the places in the cell where additional effort would make the cell smaller.

For a variety of reasons, symbolic layout editors have not replaced layout editors. Symbolic layout systems restrict designers to orthogonal layout and limited transistor shapes, which cost silicon area. Compaction can be unpredictable, so designers tend to avoid it. The kinds of operations designers perform and the kind of feedback they get with symbolic layout are different than that those with which they are familiar from exposure to layout editors. A good compromise would be a system that allowed data entry and feedback like a layout editor, but kept an internal data structure that is more symbolic (Ousterhout 1984).

Schematic Editor

A schematic editor allows a user to enter a design as an interconnection of logical blocks such as gates or transistors. The blocks are drawn symbolically and manipulated as atomic objects. They are connected with wires that indicate logical interconnection, but no physical realization. The schematic form provides a good, intuitive description of a circuit. The schematic editor facilitates entry of the drawing to generate plots for documentation or netlists for simulation and automatic placement and routing.

We can use the basic graphic editor for editing schematic diagrams. We borrow many ideas from the symbolic layout discussion:
- We define *icons*, the symbols for transistors, gates and so on.
- We let the symbolic view of the gates be the schematic representation.
- We connect gates with wires.

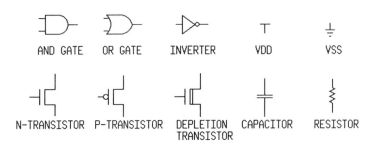

The schematic editor has only one wiring layer, but wires may cross without connecting. We start with transistor and gate primitives, but there is a nearly endless list of latches, counters, multiplexers and so on. Each primitive has a predefined icon to represent it on the screen.

We have two representations for each cell, a schematic drawing that defines the cell and an icon that represents the cell when we make an instance of it. When a user creates a new cell, he creates the icon to use to view an instance of that cell, though we can supply a default rectangular icon. Unlike layout and symbolic, the schematic does not correspond to a physical layout, so the size of an icon need not match the size of the defining cell: a counter may appear no larger than a gate, even though it is composed of dozens of gates. The icon may be customized so it represents the function better, rather than representing the implementation of that function.

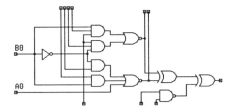

A major use of the schematic editor is to generate netlists for simulation, so it is important the the data structure map easily to the netlist. As in symbolic layout, we record connections to connectors and preserve those connections during editing. We also ensure that the connector identifiers on an icon match those on its defining schematic cell. To extract the netlist from the schematic, we follow wires to an icon indicating an instance, then continue following the wire in the defining cell of the instance.

Exercises

Programming Problems

1. Write a layout editor using the list-based object data structure. Implement boxes and instances. Include reading and writing cells.

2. Add arrays to the layout editor data structure and to your editor.

3. Add a command to add circles to the layout.

4. Define a data structure based on quad trees and incorporate it into the layout editor.

5. Define a symbolic layout data format. Write a translator from your format to layout format. What are the difficulties with translating from layout to symbolic?

Questions

1. Estimate the memory usage for the layout of a one-million-transistor chip. What problems do you foresee? Discuss possible solutions to these problems.

2. Give the order of complexity for insertion, deletion, display and moving operations for the object-based data structure. How does this compare to quad trees and corner-stitched tiles?

3. Estimate the time required to read and write the layout of a one hundred thousand transistor chip. What changes could you make to improve read and write time?

4. How does variation in the size of stored objects affect expected completion time for searches in the data structures we have discussed?

5. Estimate the time needed to display a large chip.

6. How big an area must be redisplayed when you move a box?

7. Assume all layers are displayed with transparent colors. Describe a method of redisplay that is faster than the current method. What changes to the data structures do you recommend? Is it easier to display with an object-metaphor in the user interface or with a painting metaphor?

8. Not all cells are rectangular. What changes are required to describe the boundary of a cell by a bounding rectagon? List the situations in which performance would be improved and degraded.

9. Identify the modes in your text editor. How do you know what mode you are in? How do you exit from modes? Do you find this convenient? Did you find it convenient when you were just learning the editor? Have you ever "gotten lost" in an unfamiliar program?

10. List the different criteria you would use to design a user interface for novice users and for experts. Compare these criteria against the features in Caesar and VTIlayout. Were these systems targetted to one kind of user? What must you do to make a system that caters to both?

11. How would you maintain connectivity information in a layout editor? How much computer time is required after each editing operation? What changes in data structure do you recommend? What restriction on commands do you recommend?

12. Assume that each union is a binary union: that it contains at most two other objects: unions or other objects. How many unions are needed in the data structure? If your goal is to minimize the total number of bounding box checks, including those of unions, what is the optimum number of objects in a union?

13. Estimate the storage required for the inverter in Chapter 2 stored as elements in an array, as a binary search tree, as a quad tree and as corner stitched tiles.

14. A comment in a programming language does not contribute to the function, but serves to explain the design. Describe the features of a similar kind of comment in a layout editor.

References

Layout Editor

D.G. Fairbairn and J.A. Rowson, "Icarus: An Interactive Integrated Circuit Layout Program", *Proceedings of the 15th Design Automation Conference*, 1978.

J.K. Ousterhout, "Caesar: An Interactive Editor for VLSI Layouts", *VLSI Systems Design*, Fourth Quarter 1981.

J.K. Ousterhout, "The User Interface and Implementation of an IC Layout Editor", *Transactions on Computer-Aided Design of Integrated Circuits and Systems*, v CAD-3, No. 3, July, 1984.

J.K. Ousterhout, G.T. Hamachi, R.N. Mayo, W.S. Scott, and G.S. Taylor, "Magic: A VLSI Layout System", *Proceedings of the 21st Design Automation Conference*, 1984.

VLSI Technology, "VTIlayout", VLSI Technology, Inc. 1986.

Data Structures

A. Aho, J.E. Hopcroft and J.D. Ullman, *The Design and Analysis of Computer Algorithms*, Addison-Wesley, 1974.

J.L. Bentley, "Multidimensional Binary Search Trees Used for Associative Searching", *Communications of the ACM*, vol. 18, No. 9, September 1975.

G. Kedem, "The Quad-CIF Tree: A Data Structure for Hierarchical On-Line Algorithms", *Proceedings of the 19th Design Automation Conference*, 1982.

J.K. Ousterhout, "Corner Stitching: A Data-Structuring Technique for VLSI Layout Tools", *IEEE Transactions on Computer Aided Design of Integrated Circuits and Systems*, vol. CAD-3, No. 1, January, 1984.

Symbolic Layout

C. Kingsley, "A Hierarchical, Error-Tolerant Compactor", *Proceedings of the 21st Design Automation Conference*, 1984.

Y.E. Cho, "A Subjective Review of Compaction", *22nd Design Automation Conference*, 1985.

M.Y. Hsueh, "Symbolic Layout and Compaction of Integrated Circuits", Ph.D. Thesis, University of California at Berkeley, UCB/ERL M79/80 Memo, 1979.

N. Weste, "Virtual Grid Symbolic Layout", *Proceedings of the 18th Design Automation Conference*, 1981.

J.D. Williams, "Sticks -- A New Approach to LSI Design", MS Thesis, Massachusetts Institute of Technology, 1977.

General

D.M. Meadows, J.D. Robertson, E.C. Kragh, "Computer-Aided Hybrid Microcircuit Mask Design", *International Electronic Circuit Packaging Symposium*, 1968. See also the appendix of M.D. Prince, *Interactive Graphics for Computer-Aided Design*, Addison-Wesley 1971.

W.M. Newman and R.F. Sproull, *Principles of Interactive Computer Graphics, Second Edition* McGraw Hill Book Company, 1979.

I.E. Sutherland, "SKETCHPAD: A Man-Machine Graphical Communication System", *MIT Lincoln Laboratory TR 296*, 1965.

J.D. Ullman *Computation Aspects of VLSI*, Computer Science Press, 1984.

CHAPTER 6

LAYOUT LANGUAGE

A layout language is the most versatile and ultimately the most powerful layout tool a designer can have. It gives a designer the power of a programming language to specify a chip. Many layout tasks that would otherwise require tedious placement and checking in a layout editor can be done quickly with a program. A layout language is easy to implement and is extensible by users, becoming the basis for automated layout tools such as PLA generators and silicon compilers.

A layout language attempts to describe a two-dimensional chip with a one-dimensional stream of characters. Thus the language form is hard to debug. It can be difficult to use, particularly for those who are not software-oriented, which includes most of the design community. It is a good starting point for more advanced tools, though, so you may find the layout language more valuable for you as a tool developer than it is as a layout aid in its own right.

This chapter describes an embedded language approach to developing a layout language. Such a system is very easy to implement, requiring only a few hours work to produce a minimum system. We begin by describing a simple system then a more useful one. Later in the chapter we describe useful additions. Further routes of extensibility are discussed at the end of this chapter. The layout language described in this chapter is used in Chapter 7 to develop more powerful tools.

Embedded Language

Rather than develop a new language for making layout, we *embed* layout constructs in an existing programming language following the work of Locanthi (Lang 1979), Batali and Hartheimer (1980), Karplus (1982) and others. A programming language is powerful, complete and well-defined. It has already been debugged and provides all the language features we need. It simply lacks the design features.

In this chapter we use MAINSAIL™ for our host language, though we could use any language. MAINSAIL is a Pascal-like language and the language constructs should be sufficiently familiar that they do not require explanation. To embed a layout language in MAINSAIL, we add procedures to generate layout primitives. We have already described a set of procedures to do this, the layout writing procedures from Chapter 2. In our simple system, we use those procedures directly. The result is an onionskin approach of layers of software:

design statements

lyWriter

programming language

A Simple System

We are able to write a program to generate a cell by calling the procedures in the layout writer one after the other with the proper parameters to generate the layout file. Using this method, we draw a box on the metal layer with the following code:

```
lywrite.layer(CM);
lywrite.box(1,1, 15,3);
```
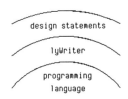

Assuming, of course, that CM is defined to be the integer we have assigned to the metal layer. We can use the same mechanism to generate polygons:

```
real array(1 to 3) x, y;
   ...
x[1] := 1;   y[1] := 1;
x[2] := 1;   y[2] := 5;
x[3] := 3;   y[3] := 4;
lywrite.layer(CP);
lywrite.polygon(x,y);
```
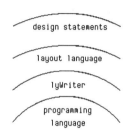

We can use similar mechanisms for generating cell headers, version numbers and so on.

A More Usable System

There are many problems with such a simple system. Frist, there is nothing to guard against users generating bad layout files. We could just execute the code above without the version number or cell header required in a layout file. We really shouldn't force our users to do the bookkeeping for building correct layout files.

Second, we have the problem of bounding boxes in cell headers. Using the model from the preceeding example, a user would have to provide the bounding box before all the data of the cell. That seems unreasonable since we could have the computer calculate the box. The problem with the computer calculating the box is that the box must be written before all the data in the cell.

The solution we use is to provide one more layer of procedures between the user and the drawing procedures. In this layer of procedures, we cache a whole cell for output, accumulating items and calculating the cell's bounding box. When the cell is finished, we write the whole cell to the output.

The layout language system includes a layout writer and procedures for the layout language. We include a layout reader so we can accept layout files from any source. The most important source will be the layout editor, but we also want to be able to build a library of cells in a layout file.

Layout Language Procedures

The section contains a list of the procedures needed to make a basic layout language. Later sections describe extensions to this language, implemented as more procedures.

File Operations

```
procedure startFile(string filename);
procedure endFile;
```

startFile opens the layout writer and writes the version number. A user doesn't have to supply the version number, we build it into the code of the layout writer. endFile closes the layout file.

Cell Operations

```
procedure startCell(string cellname);
begin
  if currentCell<>nullpointer then begin
    error("Can't start cell "&cellname&
      " because there already is a cell being defined.");
    return;
  end;
  currentCell := newCell(cellName);
end;
procedure endCell;
```

startCell does no writing. It checks that there is no cell currently being defined, then creates a new record for the current cell and sets the cell name. Just as in the layout reader, operations that create data do so in the current cell.

endCell writes the layout cell to a file. First it writes the header with the computed bounding box, then the connectors and body. The writing procedure may also optimize the data, sorting and merging adjacent rectangles.

Objects in a Cell

```
procedure box(real l,b,r,t; integer layer);
begin
  pointer(boxItem) bx;
  if currentCell = nullpointer then begin
    error("No current cell for box");
    return;
  end;
  bx := newBox(l,b,r,t,layer);
  putInList(currentCell.items,bx);
  orRectangles(currentCell.bb,bx.bb);
end;

procedure startPolygon(integer layer);
procedure polygonPoint(real x,y);
procedure endPolygon;

procedure instance(string instanceName,cellName;
  real x,y;  integer orientation);

procedure connector(string name;
  real x,y,width;  integer layer);
```

The **box** procedure puts a box into the current cell and adds the box to the bounding box of the current cell. **polygon** starts a polygon in the current cell. **polygonPoint** adds a point to the current polygon in the cell. There must be a cell and it must have a polygon open for adding points. The bounding box of the cell is also expanded to include the new polygon point.

We added a layer to the **box** and **polygon** procedures, eliminating the need for a separate procedure to set the current layer.

instance puts an instance into the current cell. The cell for this instance must already be defined. The **instance** procedure looks up the cell, transforms the bounding box of the cell as dictated by **x, y** and **orientation** and adds that box to the bounding box of the cell.

connector adds the connector to the list of connectors in the cell and expands the bounding box of the cell to include the connector.

Comments

procedure comment(string commentString);
If there is a current cell, comment puts the comment into the items list in the cell. If there is no current cell, comment writes the comment directly to the file.

Data Structures

We use the list-based data structures from Chapter 2 for our layout language. This data structure is adequate in this application since we only build the lists and traverse them sequentially to write. We do no searching or modification of the data.

The language keeps the list of defined cells, a pointer to the current cell and a pointer to a record where it accumulates the points in the current polygon.

Example

We can use our layout language to generate the inverter from Chapter 2. The inverter cell shown there is reproduced here along with the layout language code for building it.

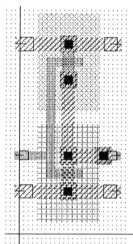

```
startFile("inv");                      # input
                                       box(-1, 12.5, 14.5, 14.5, CP);
startCell("inv");                      connector("in", 0,13.5, 2, CP);
  # the vdd wire
  box(-1, 41, 23, 44, CM);             # transistors
```

```
   connector("vddL", 0,  42.5, 3, CM);      box(6, 37.5, 14.5, 39.5, CP);
   connector("vddR", 22, 42.5, 3, CM);      box(9.5, 36.5, 12.5, 40.5, CND);
                                            box(6, 14.5, 8, 37.5, CP);
   # the vss wire                           box(9.5, 11.5, 12.5, 15.5, CPD);
   box(-1, 8, 23, 11, CM);
   connector("vssL", 0,  9.5, 3, CM);       # put the output on poly
   connector("vssR", 22, 9.5, 3, CM);       box(17, 19, 21, 19.5, CM);
                                            box(9, 16, 21, 19, CM);
   # the vdd contact                        box(17, 15.5, 21, 16, CM);
   box(9, 40.5, 13, 44.5, CND);             box(18, 16.5, 20, 18.5, CC);
   box(9, 32.5, 13, 36.5, CND);             box(17, 18.5, 21, 19.5, CP);
   box(9, 44,   13, 44.5, CM);              box(17, 16.5, 23, 18.5, CP);
   box(9, 40.5, 13, 41,   CM);              box(17, 15.5, 21, 16.5, CP);
   box(10, 41.5, 12, 43.5, CC);             connector("out", 22, 17.5, 2, CP);

   # the vss contact                        # the wells
   box(9, 15.5, 13, 19.5, CPD);             box(4, 2.5,  18, 24.5, CNW);
   box(9, 7.5,  13, 11.5, CPD);             box(4, 27.5, 18, 49.5, CPW);
   box(9, 11,   13, 11.5, CM);            endCell;
   box(9, 7.5,  13, 8,    CM);
   box(10, 8.5, 12, 10.5, CC);            endFile;

   # metal strap across the wells
   box(9,   32.5, 13,   36.5, CM);
   box(9.5, 19.5, 12.5, 32.5, CM);
   box(9,   19,   13,   19.5, CM);
   box(9,   15.5, 13,   16,   CM);
   box(10,  16.5, 12,   18.5, CC);
   box(10,  33.5, 12,   35.5, CC);
```

Symbolic Layout

New features in a layout language fit smoothly because all features are implemented as procedures. New features implemented as new procedures are indistinguishable from the original features of the language. The features we add for symbolic layout are transistors, contacts and wires.

Transistors and Contacts

We build transistor out of the layers that make it up:
```
procedure nTrans(real x,y, w,h);
begin
   box(x-w/2-2, y-h/2,    x+w/2+2, y+h/2,   CP);
   box(x-w/2,   y-h/2-1,  x+w/2,   y+h/2+1, CPD);
   box(x-w/2-4, y-h/2-4,  x+w/2+4, y+h/2+4, CNW);
end;
```

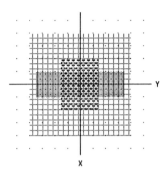

The transistor appears centered at the specified location with width and height determining the size of the contact. The transistor procedure ensures that the transistor is built correctly.

The procedure for contacts can be written to take two layers and generate the proper layout:
procedure contact(real x,y;   integer layer1,layer2);

First we check that we can legally make a contact between the two layers we are given. The definitions of legal contacts and transistors may be included with the definition of the CMOS technology in chapter 8. Many systems read a *technology definition file* or *technology file* which defines the layers, transistors and contacts as well as minimum wire widths and spacing rules for the technology.

Wires

As we discussed in our description of layout editor features, a wire is a connected series of points with a layer and width. We extend the wire construct in the layout language to include different widths and segments on different layers. We start a wire by giving an initial point, a width and a layer. We can add points or change the layer or width. Let us cover the case of all segments on the same layer first. We define procedures for creating wires that are similar to those for polygons:

procedure wireStart(real x,y,width;   integer layer);
procedure wirePoint(real x,y);
procedure wireEnd;

wireStart saves the wire width and layer in a global wire buffer. Each time we add a point, wirePoint calculates the box to be added and adds it to the current cell.

```
procedure wireStart(real x,y,width;  integer layer);
begin
  wireLayer := layer;
  wireHalfWidth := width/2;
  wireX := x;
  wireY := y;
  wireLastDirectionWasX := wireLastDirectionWasY := false;
end;

procedure wirePoint(real x,y);
begin
  real corner;
  if x=wireX then begin
    corner := if wireLastDirectionWasX then wireHalfWidth
                                       else 0.0;
    box(
      x-wireHalfWidth, ((wireY-corner) min y),
      x+wireHalfWidth, ((wireY+corner) max y), wireLayer);
    wireY := y;
    wireLastDirectionWasX := false;
    wireLastDirectionWasY := true;
  end else if y=wireY then begin
    corner := if wireLastDirectionWasY then wireHalfWidth
                                       else 0.0;
    box(
      ((wireX-corner) min x),y-wireHalfWidth,
      ((wireX+corner) max x),y+wireHalfWidth, wireLayer);
    wireX := x;
    wireLastDirectionWasX := true;
    wireLastDirectionWasY := false;
  end else begin

    ...

  end;
end;

procedure wireEnd;
begin
end;
```

We make this wire with the following code:

```
wireStart(2,2, 3,CM);
wirePoint(6,2);
wireEnd;
```

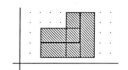

This wire drawing code only works for orthogonal wires. Extending wirePoint to work for non-orthogonal segments is left as an exercise for the reader. There is an algorithm for converting an arbitrary wire to a polygon in Mead and Conway (1980). However, it is very much to our advantage to break a wire into boxes when we can. Boxes are easier and more efficient to handle in other operations.

We find it is easy to change wire layers and widths in the layout language:

```
procedure wireNewLayer(integer layer);
begin
  if layer=wireLayer then return;
  contact(wireX,wireY, wireLayer,layer);
  wireLayer := layer;
end;

procedure wireNewWidth(real width);
begin
  wireHalfWidth := width/2;
end;
```

Frequently, we want to change just one coordinate of the wire, or we want to move relative to our current position. We define the following procedures to implement these functions:

```
procedure wireNewX(real x);
procedure wireNewY(real y);
procedure wireDX(real dx);
procedure wireDY(real dy);
```

The Example Again

Using contacts and wires, we can simplify the code for the inverter, placing wiring along a center path, terminating wires on contacts, transistors and connectors.

```
startFile("inv");

startCell("inv");
  # the vdd wire
  wireStart(-1, 42.5, 3, CM);   wireNewX(23);   wireEnd;
  connector("vddL", 0, 42.5,  3, CM);
  connector("vddR", 22, 42.5,  3, CM);

  # the vss wire
  wireStart(-1,  9.5, 3, CM);   wireNewX(23);   wireEnd;
  connector("vssL", 0, 9.5,  3, CM);
  connector("vssR", 22, 9.5,  3, CM);

  # the vdd and vss contacts
  contact(11,42.5,CM,CND);   contact(11,9.5,CM,CPD);

  # metal strap across the wells
  contact(11,17.5,CM,CPD);   contact(11,34.5,CM,CND);
  wireStart(11,16,3,CM);   wireNewY(36);   wireEnd;

  # input
  connector("in", 0, 13.5, 2, CP);
  wireStart(-1,13.5,2,CP);   wireNewX(14.5);   wireEnd;

  # transistors
  wireStart(7,12.5,2,CP);   wireNewY(39.5);
    wireNewX(14.5);   wireEnd;
  nTrans(11,13.5);
  pTrans(11,38.5);

  # put the output on poly
  wireStart(9.5,17.5, 3, CM);   wireDX(8);   wireLayer(CP);
    wireWidth(2);   wireNewX(23);   wireEnd;
  connector("out", 22, 17.5, 2, CP);

  # the wells
  box(4,  2.5, 18, 24.5, CNW);
  box(4, 27.5, 18, 49.5, CPW);
endCell;

endFile;
```

Using Bounding Boxes and Connectors

When we finish defining a cell, we write the cell to the output file and start the next one. We keep a header record containing the cell name, bounding box and connectors. We use the bounding box for computing the bounding boxes of instances, but we can also use the bounding box and connectors in the layout language to build more complex cells and to build correct layout even when we don't know in advance the sizes and locations connection points of the cells.

Bounding Box

Suppose we want to make 10 inverters, left to right, starting at $x = 7$. If we know the width of the inverter cell is 22, we loop and place each instance at $7+22i$, where i is the iteration count. Our data structures already have the bounding box of inv, so we can calculate the width of the cell and make an array of inverters without building the width of inv into the code.

```
invWidth := lookupCell("inv").bb.right-
            lookupCell("inv").bb.left;
for i := 0 upto 9 do begin
   instance("inv"&cvs(i),"inv",10+i*invWidth,0,0);
end;
```

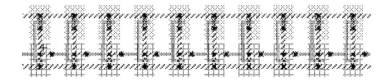

Of course, we can provide a separate procedure to retrieve the cell width.

Not only does the bounding box lookup mechanism solve the problem of determining the width of the cell, but it also continues to work after that width changes. We could write a general array building procedure that takes the name of a cell and makes a two-dimensional array of cell.

Connectors

Connectors are used to identify locations where one would like to connect to a cell. Connectors are identified by name. Just as we looked up the bounding box of the cell, we can look up a connector position inside the program, so we do not have to know connector locations in advance in order to connect to them.

Suppose, we have procedures `connectorX` and `connectorY` that look up a connector on the defining cell of the instance, then transform the x,y coordinates based on the instance x,y and `orientation`. If we wish to connect to a connector named `in` on the instance named `foo` from some `startX` and `startY`. We can write

```
wireStart(startX,startY, 2, CP);
  wireNewY(connectorY("foo","in"));
  wireNewX(connectorX("foo","in"));
wireEnd;
```

Reading Layout Files

Another essential addition to the layout language is the ability to read existing layout files. When we read a layout file, we put the cell names, bounding boxes and connectors into our `cells` list and copy the whole cell directly to the output.

This feature lets us read files written by the layout editor and assemble them in the layout language. We can also break up a chip into several layout language files, build them separately and assemble them. This feature lets us define a layout file that contains a library of cells. After reading the file, we can use any cell in the library.

Using Language Facilities

New features fit smoothly into the language as new procedures and are indistinguishable from existing features of the language. We now introduce the second powerful capability of a layout language, the ability to use expressions with existing language features. We start by writing the inverter so we can vary the size of its power and ground lines. We write the entire inverter definition as a procedure and pass the power width as a parameter. The procedure makes a new cell with its name the combination of the cell name and the power width desired.

```
procedure varyingInv(real powerWidth);
begin
  startCell("inv"&cvs(powerWidth));
  # the vdd wire
  wireStart(.5, 41+powerWidth/2, powerWidth, CM);
    wireNewX(21.5);
  wireEnd;
  connector("vddL",  0,  41+powerWidth/2, powerWidth, CM);
  connector("vddR", 22,  41+powerWidth/2, powerWidth, CM);

  # the vss wire
  wireStart(.5, 11-powerWidth/2, powerWidth, CM);
    wireNewX(21.5);
  wireEnd;
  connector("vssL",  0, 11-powerWidth/2, powerWidth, CM);
  connector("vssR", 22, 11-powerWidth/2, powerWidth, CM);

  ...

  endCell;
end;
```

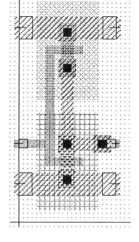

varyingInv(5.0);

We can further modify the inverter so we pass parameters to change the transistor widths for variable drive performance. That would make the cell wider (since we should preserve design rule correctness). We can even accept the drive parameter in the form of a capacitance and a desired delay to drive that capacitance, and use the programming language to calculate the necessary transistor sizes from that.

Variable sizes and positions are only one aspect of the language. We can use loops to make arrays and conditionals to create layout that appears when certain criteria are met. For example, wide power busses and greater drive on the inverter may require more contact cuts on the VDD and VSS contacts. We create those boxes only if the busses are wide enough.

Parameterized Cells

The cell `varyingInv`, is a *parameterized cell*, a cell that takes parameters as part of its definition. We can use the same inverter procedure to generate inverters with any power bus size. We can add further parameters to `varyInv` to vary transistor drive or spacing between busses. The simple inverter cell then becomes an all-purpose inverting buffer. We use the inverter with its parameters as a cell in our layout language without concern about the real cell that is written in the layout file.

We can now define a library of parameterized cells, stored as separate procedures in our programming language, that include the procedures to generate the layout. We define a parameterized cell by writing a procedure that builds its layout cell, given the parameters. The parameters to the cell are the parameters to the procedure. We select a specific parameterization of the cell by invoking the procedure with the parameters we want for our instance. The procedure generates the cell definition.

The problem with this implementation is that we must explicitly call the defining procedure with each possible set of parameters before we can make a parameterized instance. It would be much more convenient to actually keep parameterized cells and invoke instances with parameters. That invocation would generate the layout cell if it had not already been generated. The bookkeeping associated with parameterized cells would fall on the design system rather than the user of the system.

There are three ways to implement parameterized cells like these. All three require a moderate rewrite of the layout language. The major addition is a stack to store cell definitions that are not yet complete. The reason is because the user may request a specific parameterization of a cell inside the current cell being defined. If that cell has not yet been defined, we write a version of that cell with the parameters from the instance. That new cell must be written before the cell we are making. If that new parameterization also has an instance in it, the cell created for that instance must be done first and so on.

In the first method, we alter `startCell` so that instead of giving an error if there is already a cell being made, it pushes the current cell on a stack of pending cells and starts the new one. When we call `endCell`, it pops the last cell off the stack. We write a procedure for each parameterized cell and explicitly call that procedure with the parameters we want when we need to make an instance. When we encounter an instance in a cell definition, we call the procedure with the parameters, it calls `startCell` which pushes the current cell and we start defining the new one. Afterward, `endCell` pops the last cell off the stack and we return to defining it.

The second way to implement parameterized cells is to make the procedures that define cells look like cell definitions. `startCell` and `endCell` must be macros defining the beginning and end of the procedure containing the cell. Those macros also include the code to test that we have already seen a particular set of parameters for the cell. The instance statement calls the parameterized procedure for the cell before it inserts the `instanceRecord` into the current cell. Thus, all the layout code that users write is enclosed in procedures. The `endFile` calls the top-level cell definition procedure which creates all the cells with a kind of ripple effect. The programming maintains the stack of pending cells in the procedure call stack.

Rather than macros, the third method is to use procedures to explicitly build the cells, and define a parameterized cell record that includes the procedure for generating a non-parameterized layout cell, given a list of parameters, When we make an instance, we look up the cell and call its procedure with the instance's parameters to build the layout cell that we use. Because the procedure is a variable that we can find in the data structure, we can write general-purpose code to look it up. This method requires procedures as variables, which many languages do not allow.

```
procedure cellDef(
  string name;
  string procedure parameterizedCellProc(string parameters)
);
begin
  c := new(pCellClass);
  c.name := name;
  c.proc := parameterizedCellProc;
  putInList(cells,c);
end;
```

```
procedure pinstance(
  string instanceName, cellName;
  real x,y;
  integer orientation;
  optional string parameterString;
);
begin
  pointer(pCellClass) pc;
  pointer(cellClass) c;
  string nonParameterizedCellName;
  pc := lookupPCell(cellName);
  nonParameterizedCellName := pc.proc(parameterString);
  instance(instName,nonParameterizedCellname,
    x,y,orientation);
end;
```

As an alternative, we write one module for each cell. Each module contains one procedure for defining the cell and one for creating an instance. When a user requests a cell with a set of parameters, the makeInstance procedure returns a pointer to an existing cell if one has already been made or it calls the makeCell procedure to build the cell with the parameters we give it. This method is a re-packaging of the procedure method.

Of course, all these problems would have been avoided if we had defined our layout file format to accommodate cells and instances with parameters. Parameterized cells in the layout would require some way of expressing how the parameters modify the cell, so the language would have to be expanded somewhat to accept the effects of parameters. These additional features could include an entire layout language.

The Layout Language Module As a LayoutParser

We would like the layout language module to plug together with everything else we have that generates calls to a layout semantics. To do this, we define the entire layout language module as a layoutParser, so we can invoke it just like genuine layout parsers. We need not go to such extremes, though -- it is sufficient that we are able to give the layout language module a layout semantics when we run it. That way we can write to the semantics, regardless of what it is: file writer, design rule checker, or plotter.

A Procedural Netlist Generator

The usefulness of an embedded language is not limited to generation of layout files. We can write an embedded netlist language to generate schematic-style netlists for input to a simulator. The module that actually writes the file will be different, of course, as will the primitive procedures of the language. Just as with the layout language, the procedures of the netlist language allow a user to create the primitives of the netlist description: transistors, gates and wires instead of rectangles and polygons. The techniques of procedural description are universally applicable. We can quickly implement an embedded language system to generate any kind of description.

Drawbacks of a Layout Language

A layout language is not an intuitive form of description for most integrated circuit designers. It is not at all clear from the code when the design is correct, so the code must be run and the results plotted to give the kind of feedback users expect. This lengthy compile-run-plot sequence leads to a lengthy design cycle. Although the cells may be parameterized, for many applications the advantage of parameterization does not make up for the disadvantage of not being able to see the work as it is done. Typically, the parameterization features are not heavily used.

The layout language carries all the baggage of the programming language along with the power. The syntax of the language may be cumbersome for someone who does not care to use all its features. We can solve some of these problems if the language has a macro preprocessor by providing macros to relieve problems with programming language syntax.

An embedded language also reports syntax and run-time errors in terms of the native language, not the embedded one, so error messages are usually cryptic for a non-expert. One way to address this problem is to do away with the embedded language, re-implementing the language as an interpreted language. This would allow better error reporting, better syntax and a tighter design cycle, as the compile step is eliminated. It does, however, require significantly more programming to develop the parser and language semantics for all the language facilities we use. The recursive descent parser we discussed in Chapter 2 is sufficiently powerful to handle a full programming language.

Although the language form is powerful and attractive to a programmer, it is frequently unfamiliar and confusing to others. The relatively simple layout tasks require the user to manually digitize the design, converting a drawing into a list of boxes. A layout language is not a substitute for a graphical layout editor, though some of their features can be combined (Trimberger 1981, Batali, et al. 1981). Never assume your users prefer to write programs, no matter how small, no matter how simple. Such assumptions lead to "arrogant" software and will quickly relegate your work to the garbage heap.

Exercises

Programming Problems

1. Implement a procedural layout language with symbolic constructs.

2. Implement procedures and a semantics to read layout files into the layout language data structure.

3. Write procedures `connectorX` and `connectorY` that return the coordinate of the connector on an instance.

4. Write procedures `widthOf` and `heightOf` that return the size of a cell.

5. Write a procedure that breaks an arbitrary-angled wire into boxes and polygons. Pay special attention to acute angles.

6. Write a procedure that accepts the name of a cell and makes a two-dimensional array of that cell.

7. Write a procedure that takes the name of an existing instance and all the specifications for a new instance and makes the new instance that abuts the existing instance on the left. Write similar procedures to abut the new instance on the right, bottom and top.

Questions

1. How would you hide the global data of the layout language from the user? Is this an important issue?

2. How would you provide a better language? More powerful constructs? More understandable syntax? Discuss your options and the amount of work involved with each.

3. How would you keep a netlist when using the symbolic layout procedures? Discuss problems with keeping it correct.

4. Instead of breaking up contacts, transistors and wires into boxes when they are made, we could save whole contacts, transistors and wires in the list of items for the cell and translate them when we write them. What are the advantages and disadvantages of doing this?

5. What are the advantages and disadvantages of implementing contacts and transistors as cells versus procedures? In what circumstances is one method of implementation preferred over the other? In what circumstances does it not matter at all?

6. Discuss the advantages and disadvantages of implementing parameters on in the layout file format.

7. Any language with loops, conditionals and procedures is powerful enough for a layout language. Look at a specialized language (for example, knitting patterns, spreadsheet macros, picture languages or word processor macros) to see if its primitives are powerful enough to use as a layout language. Does a layout language require *all* the power of a programming language? Why or why not?

8. How would you merge the procedural concepts of a layout language with the visual capability of a layout editor?

References

Layout Languages

J. Batali and A. Hartheimer, "The Design Procedure Language Manual", AI Memo 598, Massachusetts Institute of Technology, 1980.

J. Batali, N. Mayle, H. Shrobe, G. Sussman and D. Weise, "The DPL/Daedelus Design Environment", *VLSI-81*, J.P. Gray, ed., Academic Press, New York, 1981.

K. Karplus, "CHISEL, An Extension to the Programming Language C for VLSI Layout", Ph.D. Thesis, Department of Computer Science, Stanford University, 1982

C.R. Lang, "LAP User's Manual", Computer Science Department Technical Report #3356, California Institute of Technology, 1979.

S. Trimberger, "Combining Graphics and a Layout Language in a Single Interactive System", *Proceedings of the 18th Design Automation Conference*, 1981.

MAINSAIL is a registered trademark of Xidak, Inc.

CHAPTER 7

LAYOUT GENERATORS

A *layout generator*, sometimes called a *cell compiler*, is a program that generates layout for a function using a predefined structure. The range of software that falls into this definition of layout generator is very large, encompassing everything from the parameterized inverter in Chapter 6 to placement and routing. We take a bottom-up approach to layout generators, building on our layout language and developing larger and more software-intensive generators.

This chapter starts with a discussion of ways to build a parameterized NAND gate. The goal of this first section is to describe several successful techniques for building layout generators. The chapter continues with more software-intensive procedures to generate large structures such as PLAs. Then we introduce silicon compilation and placement and routing.

Parameterized Cells

Let us continue our discussion of parameterized cells from Chapter 6. Suppose we have a two-input NAND gate and a four-input NAND gate:

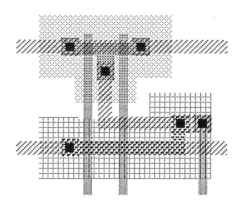

Two-input NAND

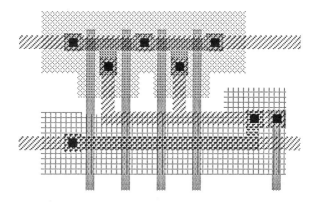

Four-input NAND

We want an n-input NAND, where n is a parameter we pass to the NAND procedure. For simplicity, we will restrict n to be even for now. An n-bit NAND gate should look like this:

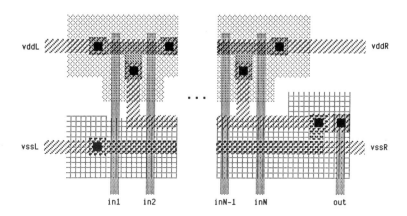

There are two basic methods for making the NAND gate, a *language-based* method and an *instance-based method*. We will discuss both methods plus techniques that combine the two.

Language-Based Method

The language based method is an extension of the techniques we used in Chapter 6. We insert more and more language constructs to build the parameterization we want (Ayres 1983). To add two more bits to a NAND gate, we add code to make the new contact and input polysilicon lines, and to extend the transistor structures horizontally. There are two parts to this task, making the loop that builds the layout and placing the parts of the layout on the right side of the cell.

The code for a variable-sized NAND gate follows. The length of the VDD and VSS wires and the X position of the right side VDD and VSS connectors depend directly on the number of inputs. We use an equation to calculate the right side of the cell for those positions and we build a loop that implements two bits of the NAND gate each time through the loop, placing the input lines, the output contact and the VDD contact. We build the layout for the output after we've drawn all the input lines.

```
startCell("nand"&cvs(nInputs));

# vdd wire
connector("vddl", 0.,32., 3., CM);
connector("vddr", nInputs*8.+28.5,32., 3., CM);
wireStart(-1.5,32., 3., CM);
  wireDX(nInputs*8.+32.5);
wireEnd;

# vss wire
connector("vssl", 0.,9.5, 3., CM);
connector("vssr", nInputs*8.+28.5,9.5, 3., CM);
wireStart(-1.5,9.5, 3., CM);
  wireDX(nInputs*8.+32.5);
wireEnd;

# the p channel transistor diffusions
contact(10.5,32., CM,CND);
wireStart(10.5,31.5, 3., CND);
  wireDX(nInputs*8.-1.);
wireEnd;

# the n channel transistor diffusions
contact(10.5,9.5, CM,CPD);
wireStart(10.5,9.5, 3., CPD);
  wireNewX(nInputs*8.+19.);  wireDY(5.5);
  wireDX(.5);  wireNewLayer(CM);
wireEnd;

# loop to make a pair of inputs
for i := 1 upto nInputs div 2 do begin
  basex := 14.5+cvr(i-1)*16.;   # the center of the first
  # first input
  wireStart(basex,-1., 2., CP); wireNewY(35.); wireEnd;
  connector("IN"&cvs(2*(i-1)+1), basex,0., 2., CP);
  # second input
  wireStart(basex+8.,-1., 2., CP); wireNewY(35.); wireEnd;
  connector("IN"&cvs(2*(i-1)+2), basex+8.,0., 2., CP);

  # the output contact for the bits
  wireStart(basex+4.,31., 4., CND);
    wireNewY(27.5);  wireNewLayer(CM);
    wireNewWidth(3.);  wireNewY(13.5);
  wireEnd;
```

```
  contact(basex+12.,32., CM,CND);    # the VDD contact
end;

# the output wire
wireStart(17.,15., 3., CM);  wireNewX(nInputs*8.+24.5);
  wireNewLayer(CP);  wireNewY(-1.);  wireEnd;
connector("out", nInputs*8.+24.5,0., 2., CP);

# the wells
box(3.5,2.5, nInputs*8.+26.5,16.5, CNW);
box(nInputs*8.+12.5,16.5, nInputs*8.+26.5,22., CNW);
box(3.5,25., nInputs*8.+17.5,39., CPW);
box(11.5,19.5, nInputs*8.+9.5,25., CPW);

endCell;
```

Instance-based Method

A major difficulty with the language-based method is the language itself. If you had read and understood the code of the NAND gate as you should have, you would have a good intuitive grasp of the difficulty of understanding the language method. Much of the code of the NAND is actually graphical objects placed on layers at set locations, or locations relative to offsets. The language form is a very poor form from which to visualize the layout. The easiest way to understand the language description of the cell is to draw an example cell on graph paper, simulating execution of the code. We now explore a much simpler instance-based method where we use a layout editor to make the graphical pieces and only use the language to assemble them (Martínez and Nance 1984, Edgington et al. 1984).

We can define the *n*-bit NAND in terms of three cells: one for the left side of the gate, before the input lines; one for the input lines; and one for the output on the right side of the gate. We build an *n*-input gate by replicating the middle cell for the number of bits we want.

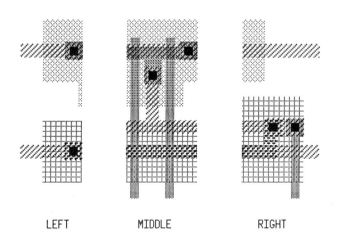

```
         LEFT            MIDDLE          RIGHT
```

```
instance("left","leftcell",0,0);
x := widthOf(leftcell);
for i := 1 upto nInputs div 2 do begin
  instance("mid"&cvs(i),"middleCell",x,0);
  x := x + widthOf("middleCell");
end;
instance("right","rightcell",x,0);
```

Of course, we don't actually save any work, because each cell must still be defined, but this instance method is easier to read and easier to debug. The individual cells contain all the layout, but have no parameters and can be laid out with a layout editor. The language portion creates no layout primitives, it just assembles the pieces. Thus, both the cell design and the coding for the parameters are easier and more understandable (Rowson 1980).

To allow an odd number of bits, we can define another cell for the right side that includes the last odd bit. That cell consists of the left side of the two-bit middle cell plus the right side cell. The code that builds the NAND gate chooses the even RIGHT or the odd RIGHT depending on the number of bits required.

As an alternative, we can define a cell for a single bit as shown below and replicate it, mirroring every second bit and placing them so the contacts overlap. The right side cell will still differ depending on whether the number of bits is even or odd. This technique requires more careful programming, but requires less cell design.

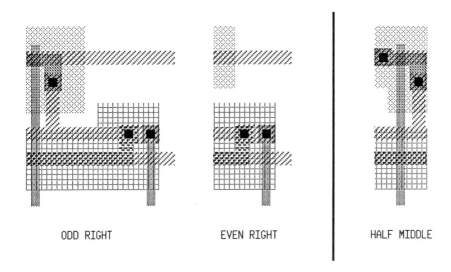

ODD RIGHT   EVEN RIGHT   HALF MIDDLE

Suppose we want to vary the width of the transistors in the NAND gate. The layout of the cells is fixed in the instance method, but we can build *instance overlays* to program the cells. An instance overlay is placed at the same location as the instance and contains only the special programming needed for that cell. We can make several overlays of transistors of various sizes and choose the one we want depending on the amount of output drive we want from the gate. We can even change the programming on a bit-by-bit basis. We wind up placing two cells at each location for the bits in the gate, one for the common "base", one for the programming.

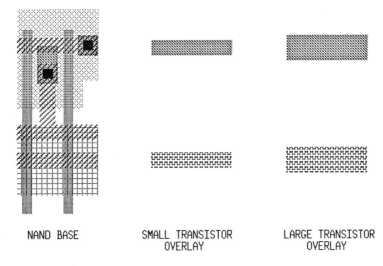

NAND BASE   SMALL TRANSISTOR OVERLAY   LARGE TRANSISTOR OVERLAY

Look closely at the middle cell we defined originally. The metal line for the output extends all the way across the middle cell. It is necessary that the line cross the cell so the third and fourth bits connect to the first and second. However, this cell leaves a small unnecessary tab of metal to the left of the first bit:

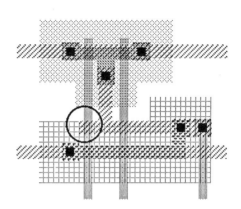

If we want to remove that bit of metal, we can define a new cell for the leftmost middle cell or we can leave the output line out of the middle cell completely, then draw the metal line over the cells when we are finished:

```
instance("left","leftcell",0,0);
x := widthOf(leftcell);
for i := 1 upto nInputs div 2 do begin
   instance("mid"&cvs(i),"middleCell",x,0);
   x .+widthOf("middleCell");
end;
instance("right","rightcell",x,0);
wireStart(widthof("leftcell")+6,17,3,CM);
   wirePoint(x,17);
wireEnd;
```

This technique of building a base of identical cells then drawing over them with programming is a powerful one and can be used in many situations. We see it again in the PLA generator.

PLA Generator

A programmable logic array (PLA) is a regular structure for laying out arbitrary logic functions. The PLA implements two-level AND-OR logic, thereby implementing any boolean logic function. The PLA structure is divided into six major parts, shown in the following figure: input buffers, AND plane, OR plane, output buffers and two pullup blocks.

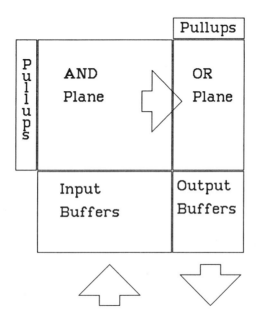

The PLA is an extensible structure. Any number of inputs may come in at the lower left and into the AND plane. Each row of the AND plane is one gate forming one *product term* or *minterm*, one AND of any number of the input lines. These product terms pass across to the OR plane where each column is one gate. The OR plane forms sums of the product terms and that result is passed to the output.

The following figure shows the transistor representation of an NMOS PLA. The inputs are driven into the array in their true and complement form, so any combination of either sense of the inputs can be calculated. The AND-OR array is implemented as INVERT-NOR-NOR-INVERT, for efficiency. Therefore, the inverted input is used as the true term in the calculation, and the outputs are inverted to give their correct sense. The inclusion of a signal in a gate, either in the AND-plane or the OR-plane is done with a single transistor.

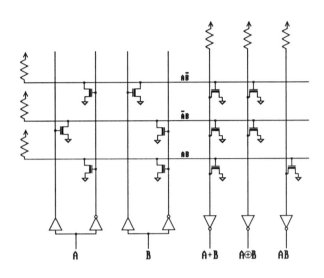

PLA Assembly

The main advantage of the PLA is that it has a regular, repeated structure. The input and output sections are repeated structures, like the input part of the NAND gate we just examined. The same is true for the pullup circuitry on the left side of the AND plane and on the top of the OR plane.

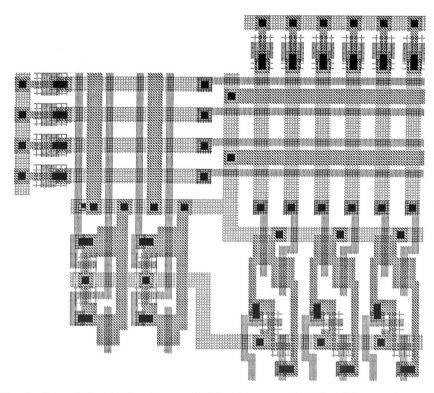

The bodies of the AND plane and OR plane are slightly different, because not all elements are the same. Although we can divide the AND plane and the OR plane into grids of cells, there are two different cell types, corresponding to inclusion of the term in the gate and exclusion of the term from the gate. Since programming transistors share ground lines, cells at odd-numbered locations must be mirrored so they connect correctly. We can use the instance method to build a PLA, making a two-dimensional array of cells, picking either the **include** cell or the **exclude** cell for each location in the array, mirroring each one appropriately for its position.

However, the PLA naturally divides into two-bit pieces: each input is driven into the array in its *true* and *complement* form; in order for the outputs to fit into the narrow pitch, they are drawn two to a cell; pairs of cells in the AND and OR planes share ground wires. Therefore, we make our job easier if we divide the PLA along 2-bit lines, as we did with the NAND gate. We can build the PLA with these cells:

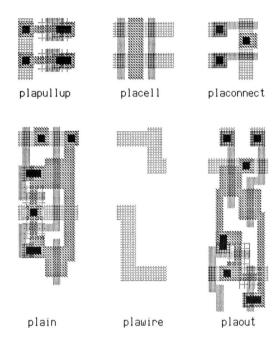

plapullup   placell   placonnect

plain   plawire   plaout

We assemble these cells to get a blank PLA like the one on page 151. After building the blank array, we draw the programming over the cells to *personalize* the PLA. plapullup, placonnect, plain and plaout each have connections for two bits. placell has connections for four bits: two product terms on each of two input lines. As we discussed in the last section, we can take care of the case when we have odd numbers of bits by defining special single-bit end cells.

Layout Generators

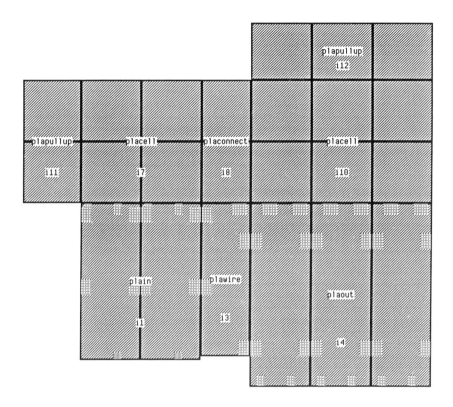

PLA Code File

Before we go any further, let us re-examine our problem and our potential solution. If we continue to work as we have through the last chapter and this one, our next step will be to write a parameterized cell that takes PLA code and generates the layout. Instead, let us write a general PLA builder that accepts a *PLA code file* as input and generates the layout.

We have many reasons for making a special PLA code file. First, it is easier for people to use than the layout language. A user can type a file that represents the PLA as a two dimensional array rather than building a PLA with procedure calls in the layout language. With a translator from the PLA code file to layout, we can guarantee that the entire structure is built correctly every time. Further, we have a data format for the logic in the code file independent of the implementation. We can optimize the code file logic to make better PLAs without having to worry about the implementation of that PLA (Brayton 1984). We choose a PLA file format with this BNF:

codeFile	= { comment } sizeLine { codeLine }
	{ connectorLine r.tt fl
comment	= # any characters <eol> r
sizeLine	= Ninputs Noutputs NproductTerms r
codeLine	= comment | codeBodyLine r
codeBodyLine	= { term } <eol> r
term	= "0" | "1" | "-" r
connectorLine	= { string } <eol> r

The code file is split into lines. The first non-comment line indicates the size of the PLA: the number of inputs, outputs and product terms. In the body of the PLA, "0" indicates a connection to the *false*-line in the AND plane and no connection in the OR plane, "1" indicates a connection to the *true*-line in the AND plane or to the product term in the OR plane, and "-" indicates no connection in the AND plane and don't-care in the OR plane.

There is one code line for each product term. The number of terms on a line is equal to the sum of the inputs and outputs. Blanks are allowed in the code lines to improve legibility. The connector lines give names to the connectors starting at the leftmost input and ending at the rightmost output. The following is a sample code file:

```
# code file for one-bit full adder
3 2 7
#inputs    a b cIn
#outputs   sum cOut

100 10
010 10
001 10
111 10
11- 01
1-1 01
-11 01
a b cIn    sum cOut
```

This PLA represents the equations:

```
sum  = a -b -cIn + -a b -cIn + -a -b cIn + a b cIn
cOut = a b + b cIn + a cIn
```

Constructing a PLA Generator

We write a parser to read PLA code files and build the PLA. We embed the PLA generator in our layout language to allow others to use it either from the language or as a stand-alone translator. We build a PLA by first constructing a blank array, then programming the array as we read the body of the file.

```
read size line of code file

# build the blank base
y := 0;
for j := 1 upto productTerms div 2 do begin
  x := 0;

  for i := 1 upto nInputs do begin
    instance("A"&cvs(i)&cvs(j),"placell",x,y);
    x := x + widthOf("placell");
  end;

  instance("b"&cvs(j),"plaConnect",x,y);
  x := x + widthOf("plaConnect");

  for i := 1 upto nOutputs div 2 do begin
    instance("O"&cvs(i)&cvs(j),"placell",x,y,rotate270);
    x := x + heightOf("placell");
  end;

  y := y - heightOf("placell");
end;

# now program the array
y := topBit;
for j := 1 upto productTerms do begin
  x := 0;
  for i := 1 upto nInputs do begin
    t := nextTerm;
    case t of begin
      ['1'] begin
        contact(x+rightbit,y,CM,CD);
        wire(x+rightbit,y,4,CD);  wireDX(-6);  endWire;
      end;
      ['0'] begin
        contact(x+leftbit,y,CM,CD);
        wire(x+leftbit,y,4,CD);  wireDX(6);  endWire;
```

```
        end;
        ['-'] begin
          # not included, do nothing
        end;
      end;
      x := x + widthOf("placell");
    end;

    x := x + widthOf("plaConnect");

    for i := 1 upto nOutputs do begin
      ...
      x := x + heightOf("placell") div 2;
    end;

    y := y - heightOf("placell") div 2;
  end;
```

We build the PLA core array starting at 0,0 and working down and to the right. `leftBit` and `rightBit` are constants that indicate the displacement of the programming points in the `plaCell` from the left edge of the cell. Similarly, `topBit` gives the distance from the bottom of the cell to the top programming point in `plaCell` (`plaCell` contains four programming points: two product terms in each of two inputs). We program the bits in the AND plane in pairs, since the code file indicates whether to include the input bit *true* or *false*. However, we program the OR-plane bits individually since each column corresponds to one output.

Because the OR-plane cells are rotated with respect to the AND-plane cells, the programming must be rotated also. Although we move `x` by the width of the `placell` and `y` by the height, the two must be equal, since the cells are rotated in the OR plane and must maintain a straight line.

The PLA As a State Machine

A *state machine* is a device that has a state, inputs and outputs. The outputs of the device and its new state depend on the inputs and the current state. We build a state machine with a PLA by routing outputs back to the inputs. The recirculated outputs, known as *feedbacks* constitute the state of the machine. Each clock step, the PLA determines its new state and outputs based on its current state and inputs. The PLA implements any such state machine, making it a very powerful general structure for control applications such as a micro sequencer.

Our PLA generator should have the option to generate the feedback wiring through clocking transistors, given the number of feedbacks in the state. Other inputs and outputs of the PLA may also be gated with the clock or left asynchronous.

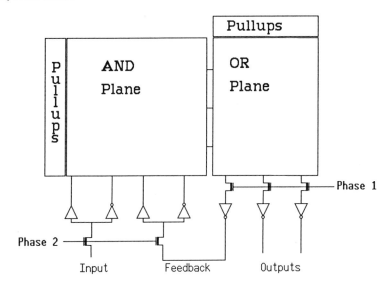

The Periphery

The PLA core is surrounded by other circuitry. On the left and top are pullup circuits. On the bottom are input drivers and output buffers. These pieces are regular arrays and are straightforward to build by tiling with regularly-spaced instances. But around the periphery of the core are other support circuits that we cannot do so easily, such as feedback routing on the bottom and wider power connections on all sides. The requirement for better ground connections causes irregularities in the instance tiling loops. These peripheral issues involve more language-style development and account for the bulk of the coding and debugging time.

Additional Power and Ground

When a PLA contains a large number of product terms, each ground wire in the AND plane is required to sink a large amount of current. To ensure proper grounding, we add additional ground connections to those ground wires. These connections come from the **placonnect** between the planes and we insert them between two product terms. We enlarge **placonnect** cells to contain a wider ground line and we include the ground connection in the AND plane and the spacing for it in the OR plane and pullups.

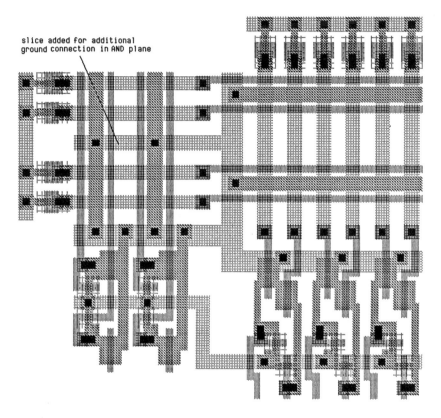

It is easier to supply more current on the VDD line since the core contains no VDD connections. We can place those connections around the periphery by substituting input buffers, output buffers and pullup cells with larger VDD wires.

Finally, large PLAs require larger input buffers, so we make cells for the input drivers that take a drive capability parameter. We can program the drivers with overlays, like we built the variable-power NAND, or we can have several different buffer cells and select the correct buffer when we read the PLA size line.

Introduction to Silicon Compilation

A *silicon compiler* is a program that translates a high-level description of a function into the layout needed to fabricate an integrated circuit to perform that function. This definition admits many systems that are not generally recognized as being silicon compilers, such as placement and routing systems, the PLA generator we just built and even the n-input NAND gate. The term *silicon compiler* has been used as a name for these types of tools. However, we further restrict the definition of silicon compiler, as many authors do, with the requirement that the layout produced by the compiler is comparable in density with that produced by an expert human layout designer. This restriction limits silicon compilers to the kinds of array structures that give high density, usually memories and computer data paths.

In this section, we investigate the problem of generating the layout for a multiple-bit processor datapath. In Chapter 9 we discuss functional models for large blocks as well as methods for developing test sets, other steps in the silicon compilation process.

Datapath Compiler

A *datapath* is the data processing section of a processor. It consists of several multiple-bit *data path elements* or *operators* arranged horizontally and connected with busses. Control signals connect to the datapath at the top and bottom. The datapath is a canonical structure that we can lay out automatically.

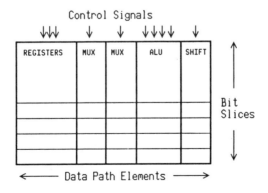

A Datapath Element

A datapath element may be an ALU, multiplexer, register file, shifter or other function that operates on multiple-bit data. Each data path element consists of a cell for a single bit of that element replicated vertically for the number of bits in the data path. The wires that make up a bus are spread among the bits of the data path elements. One data path element for a 4-bit datapath looks like this:

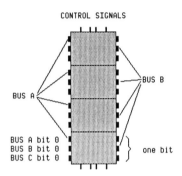

We define datapath element bit cells so they are all the same height and, when they are stacked vertically, their control signals connect vertically between bits in the bitslice. We build a datapath element by making a one-dimensional array of bit cells. We place data path elements next to one another to build the completed datapath. The datapath may be assembled by tiling with fixed, drawn instances.

However, we cannot replicate a single cell to build a datapath element because not all bit positions are the same. An ALU may contain carry lookahead circuitry that is dependent on the bit position. We may want to build different cells for even and odd bit positions so we can share wells in CMOS. Each bit of each datapath element must be built specifically for its bit position. Viewing the data path elements as parameterized cells, the datapath construction code looks like the following code. The procedure to build a datapath element cell may actually construct the cell or it may select a different cell depending on the bit position. The distinction between the code of the compiler and the code of a parameterized cell is not very great.

```
y := 0;
for bitpos := 0 upto nBits-1 do begin
  x := 0;
  for all data path elements (e) do begin
    cell := selectDPEcell(e, bitpos,nBits-1);
```

```
    makeInstance(cell,x,y);
    x := x + widthOf(cell);
  end;
  y := y + heightOf(cell);
end;
```

Connecting the Busses

All bits of a data path element are connected identically to the busses. We can investigate the wiring of one bit of the datapath and duplicate the wiring for all bits to make the busses. We examine two methods for connecting the busses. Each has advantages and disadvantages.

We can define a bit cell so the busses run across the cell and connect by abutment. The busses need not cross the cell on a single layer or in a single straight line, they may be laid out in any fashion that is reasonable for the layout of the cell, as long as they have connectors at the predefined positions on the left and right. We pass parameters to the cell indicating to which busses the cell must connect. The cell layout must be parameterized to connect any input and output to any bus. When we indicate that the cell input is to be connected to the top bus, the code that builds the cell inserts the contact to that bus.

With this organization, we can use the instance method to lay the suitably-parameterized instances side-by-side to make a bit slice and replicate that bit slice to make the full width of the datapath.

Although the assembly is simple, the cell layout is complicated. Since the inputs for a datapath element may be on any of the busses, we must design the cells so they can be parameterized to connect inputs and outputs to any of the busses. This feature requires extensive use of language constructs to build the cell. Alternatively, we could design the cells with multiplexers and three-state drivers to connect to the busses and let the user decide by the control signals which element connects to which bus. This option makes the cells much larger. The user may wish to break some busses to have separate communication on different segments of a bus. We want to break busses inside a cell, so we have further parameters to the cells. Coding and maintenance of these complicated cells is difficult.

We can apply some restrictions to the busses to reduce the number of options in the cell and reduce the amount of code needed. For example, if we have four busses, we can define the data path cells so they always break two of the busses, which we call *local* busses, and never break the other two, *global* busses. Each datapath element can connect to either the local bus or the global one. We make an ALU with multiplexers that allow it to take each of its inputs from one local-global set of busses on one phase of a clock and output to either global bus on a different phase of the clock. We trade busses for time multiplexing. We can make similar restrictions for all cells, as was done by Johannsen (1981).

This wiring method works well when there is only one metal wiring layer, since the busses must be wired in that layer and the cells must also use that layer. However, it requires special "bus break" cells to separate pieces of the global busses that should not be connected. It is also difficult to extend to more busses than fit across the height of the cell. The major drawback of this wiring method is that it requires a large amount of language-style code.

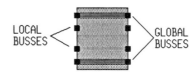

The second method of making busses lays the busses on top of the cells (Rowson et al. 1987). This is the instance method with overlaid programming, just as we saw in the PLA generator. The cells must be designed with open space for all busses, but this is not a serious limitation in a design technology that has several layers of metal: the designer avoids using the bus wiring layer in the cell, and we use only that layer for wiring the busses. The cells are simpler, they don't have to make the connections to busses, but the cell must be designed so we can connect to any bus easily. We do this with connectors that extend the entire height of the cell. No matter where we put the bus vertically across the cell, we can drop a contact at the connector position to make the connection.

This technique requires a wiring layer dedicated to bus routing and restricted use of the second wiring layer for the connectors. The cell design may not be as efficient, since the vertical connection bar for each signal takes a lot of space. But it is easy to add additional space for busses, we just extend the connection bars outside the cell to connect to busses that do not fit over the cell. In addition, this method allows us to easily optimize placement of the elements.

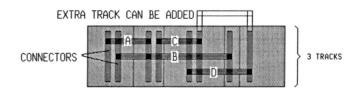

Optimization of the Datapath

We make a bus by assigning it to a track across the data path elements. The *span* of a bus is the range from its leftmost connector to its rightmost connector. Two busses may share the same bus track across the cells if their spans do not overlap. In the preceeding figure, bus B may not share a track with any other bus. Bus A may share a track with C or D. The order in which the datapath elements are placed determines how many tracks we require to connect the cells. There are a limited number of tracks that can fit over the cells. If our design requires more, we extend the tracks off the top of the cells and the resulting datapath layout will be larger.

We can require a user to specify the order of the datapath elements and the assignment of the busses to bus tracks. This solution works with small datapaths and small numbers of busses, but large datapaths may be difficult for a user to optimize. Alternatively, we can attempt to re-order the datapath elements to minimize the number of tracks.

We can calculate the total number of bus tracks required at each data path element location, assuming the maximum sharing of tracks. This number is called the track *density*. In the following figure (a), the density is three, since A, B and D all must be wired past the third element. However, if we swap the positions of the third and fourth elements, we can lower the density to two, seen in figure (b).

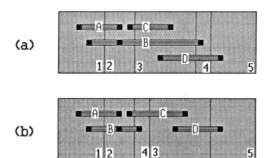

To optimize the datapath, we start with an initial ordering and perform *placement improvement* by swapping positions of pairs of datapath elements to find a lower density. If we find a lower density, we change the order of the data path elements and try again. Since there are not many elements in a datapath, we can try all exchanges. We quit trying when we fail to find an exchange that makes a better track assignment. This operation is called *pairwise interchange*. When pairwise interchange finishes, we assign busses to tracks, add the busses as overlays on the datapath elements and add the contacts to the connectors.

Specifying a Datapath

The most straightforward way to describe a datapath is by specifying the number of bits of the data path and listing the cells in order and the busses to which they connect. We can describe a datapath with a small datapath language:

```
bits 16;

port LEFT IN
register 4 R1 R2 OUT
mux IN R1 A
mux OUT R2 B
alu A B C
shift C OUT
port RIGHT OUT
```

This description makes a data path 16 bits wide, with bus IN entering from the left side, a bank of four registers that can be written from the OUT bus and which write to R1 and R2. The ALU takes input from the multiplexers that choose the input from the registers, the input or the output of the shifter. The ALU output goes through the shifter and goes out the right side, into the ALU or into the registers. LEFT and RIGHT indicate the side of the datapath that the bus enters or leaves. **bits** gives the word size of the datapath. The other names are the names of data path elements in our library. We write code to take this description and generate the layout by placing pre-defined cells for the data path bits. We build the data path incrementally from the bits.

Just as with a layout language, a datapath language is harder to understand and debug than a picture. We can specify a datapath in a more convenient fashion as a schematic drawing using icons to represent the datapath elements and a separate icon to express the word size. In essence, it is a drawing of one bit slice of the datapath. This drawing is the same datapath specification we saw in the language form, and describes connections, but not the order of the elements in the datapath. The system may choose an order to minimize bus congestion and wire length.

Layout Generators

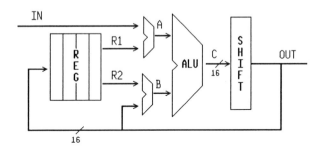

Parameters On the Data Path Elements

It is tedious for a user to specify each register separately. In addition, a block of registers can share circuitry, so we can make a better layout if we allow a user to include a block of registers instead of single registers. The register code takes a parameter that determines the number of words of memory and replicates the memory cell, sharing the overhead. We allow a user to specify parameters on cells, such as the number of words in a register or the speed of an ALU.

We can implement these functions as parameterized cells or as separate cells. If we choose to implement them as separate cells, we can hide the separation from the user, letting him assign parameters to cells in the specification, or we can make them visible, requiring him to select different data path elements for the different options. A successful system will do some of both. It may be more effective to allow a user to specify the number of words in a register as a parameter, but to select a different ALU if he wants high performance or a different carry structure. In this list, the cells are listed as functions with parameters, without regard to the way we choose to implement them.

- ALU
  - speed
  - ripple/carry lookahead
  - increment, decrement only
- register
  - number of words
  - multiple port options
  - LIFO, FIFO, CAM, Read-Only with value
- multiplexer
  - number of inputs
- shifter
  - number of bits
  - rotate

- IO port
  - drive capacity
- test structures
  - signature generator
  - scan path register
- multiplier
  - number of bits in multiplicand
- zero detect
- counter

The Periphery

The control signals in the datapath must be driven sufficiently to switch all the bits in the datapath. Rather than require a user to build sufficiently-large buffers, we can make them ourselves and wire them to the top of the datapath. Since the datapath control signals are not spaced evenly, this is not a simple array that we can abut to the top of the datapath. We may have to *river route* the two pieces together. River routing is the subject of one of the assignments.

VDD and VSS are connected to the datapath horizontally though the data path cells. Users connect to VDD and VSS on the sides of the datapath. If the datapath is very wide, we will be feeding insufficient power into the central data path elements, so we insert a datapath element to refresh the power connections.

Not all datapaths are as simple as we have described. Additional features such as inter-bitslice wiring, sections of the datapath with fewer bits and signals brought into the datapath from the top or bottom complicate the assembly. We will not discuss the problems involved with these features.

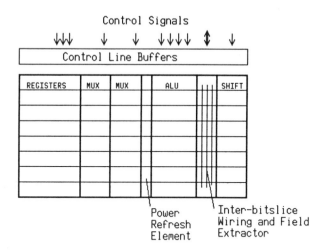

A Small Matter of Circuit Design and Layout

We have ignored the major part of datapath design: the design and layout of the cells. There are two main reasons for this. First, a detailed description of the layout of the cells and the design techniques is well beyond the scope of this book. Descriptions of data paths have been given by Johannsen (1981), Rowson et al. (1987) and others. The major considerations in design of the datapath cells are the number of busses that may cross a cell, the height of the cell, a mirroring scheme, performance of the resulting datapath and a clocking style. The assembly changes for these decisions are minor.

The second reason for us not pursuing the cell design is that the code of our compiler should not be dependent on the details of the cell implementation. We define the physical and structural interface for the datapath cells, as we did at the start of this section. The cell design must conform to that interface and the assembly code must rely on the interface being met. The full layout is not important for correct operation of the code that assembles it.

Introduction to Placement and Routing

The most common technique for automatic layout generation is *placement and routing*. A placement and routing system uses a predefined floorplan to build layout from a netlist of predefined cells. The floorplan has areas filled with logic, the *cells*, and areas for routing signals among the cells, the *channels*. We divide the physical design task into three phases. During *placement*, we assign cells in the netlist to physical locations in the resulting chip. During *global routing*, we choose the routing channels through which each wire will pass. During *local routing*, we complete the physical realization of the wires.

There are three major types of floorplans for automatic physical design using placement and routing: *standard cell, gate array* and *arbitrary block*. In a standard cell system, the cells represent logic gates such as AND, OR and INVERT, but may range up to latches and counters. The cells are assembled into rows and separated by variable-sized routing channels, where we make the connections among the connectors. A gate array *base* consists of a prefabricated array of transistors. Usually only the metal wiring layers are left undefined. The cells in a gate array are the same complexity as those in standard cells and are defined by customizing a few transistors with a small piece of metal wiring. The gates are arranged in rows like the standard cells, but the routing channels between rows in a gate array cannot expand because the locations of transistors are fixed. In an arbitrary block system, the cells can be any size and routing channels fill the spaces between them. The arbitrary block system attempts to place cells to make a rectangular chip while balancing the need for short wires.

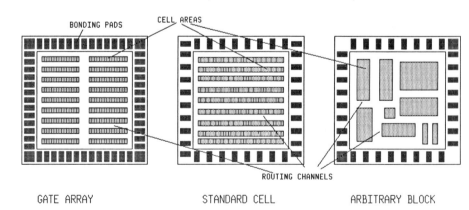

GATE ARRAY   STANDARD CELL   ARBITRARY BLOCK

Throughout this section, we refer to wires as *nets* and instances as *cells*, in accordance with placement and routing literature.

Standard Cells

This section describes a placement and routing system for standard cell chips. It is intended as an introduction to the techniques and issues of placement and routing for all kinds of systems, but it is not comprehensive. The reader is directed to the survey by Soukup (1981) and the other references for more detail.

In a standard cell system, we define cells that perform the functions of logic gates. The cells are rectangular and are all the same height. They are designed so VDD and VSS connect to adjacent cells on the left and right and external signals come out the top and bottom. To place a set of standard cells, we determine the order of the cells and line them in rows on the chip, connecting the power supplies. We wire between them with two wiring layers in *routing channels* between rows of standard cells. We make the routing channel as wide as we need to make the connections and adjust the placement of rows to minimize the vertical space.

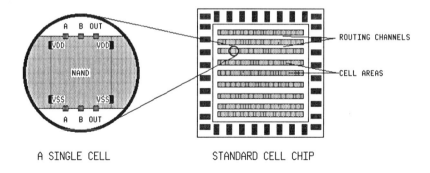

A SINGLE CELL   STANDARD CELL CHIP

Important Concepts

A routing *channel* is a rectangular area with connectors on the top and bottom. A *track* is a horizontal slice along the channel that is wide enough for one wire. We assume that tracks are spaced far enough apart that we can put contacts on adjacent tracks without violating a design rule. We measure the congestion of a channel at any point as the *density*, the number of nets that cross a vertical line across the channel at that point. The maximum density in a channel is the minimum number of tracks that we can use to route the channel, so it gives a lower bound on the height of the channel. If we minimize the *channel density*, we minimize the height of the channel.

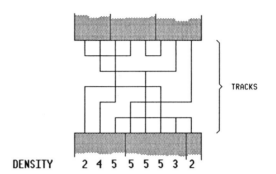

Placement

Although the bulk of the space on chips built with placement and routing systems is taken up with wiring, the key to efficient wiring is good placement. The goal of placement is to find an order of the cells so the resulting chip is small and fast. Overall chip size depends on minimizing wiring congestion; overall chip speed depends on the length (and hence the capacitance) of wires on the critical timing path. It is the goal of the placer to assign cells to positions on the chip so as to optimize these two possibly-conflicting goals. Finding an optimal placement has been shown to be NP complete, but we do not require an optimum placement, merely a good one, so we can use some good heuristic algorithms.

We divide placement into two stages, *initial placement* and *placement improvement*. A powerful algorithm for initial placement is *min-cut partitioning* initially described by Kernighan and Lin (1970), and later refined by Breuer (1977), Dunlop and Kernighan (1985) and others.

The goal of min-cut partitioning is to split the netlist into two smaller netlists, minimizing the interconnection between them. We recursively apply min-cut partitioning, splitting the sub-netlists into smaller subsets until we have partitioned the netlist into sets of only a few cells each. Conceptually, we partition the area of the chip horizontally and vertically to determine the final locations of the cells. When we are finished splitting sets of cells, we form the cells into rows, choosing the row and the order in the row based on the subset in which they finally reside.

The min-cut algorithm separates the netlist into two sets arbitrarily, then exchanges cells between the sets if such an exchange reduces the *cut count*, the number of connections between the sets. Exchanges continue until no exchange reduces the cut count.

We call the two sets A and B. If a is a cell in set A and it is connected to a net that has all its other connections in set B, we can decrease the cut count by 1 if we move cell a into set B. We call E_a the number of nets connected to cell a for which cell a is the only cell in set A. This is the number of crossings we would eliminate by moving cell a into set B. If cell a is connected to a net that is wholly contained in set A, then moving a into B would increase the cut count by 1. We call I_a the number of nets connected to cell a that have no connections outside set A. The gain for moving cell a into set B is $D_a = E_a - I_a$. The gain for exchanging cell a in set A for cell b in set B is $gain = D_a + D_b - c_{ab}$ where c_{ab} is the number of nets that contribute to both E_a and E_b. It compensates for cells a and b being on some same nets.

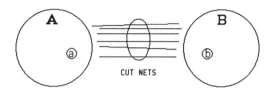

One possible implementation is a *greedy* algorithm. We find the pair of cells in A and B that give the greatest gain and exchange them. We continue, taking every exchange that gives an improvement in the cut count. However, cells tend to be parts of heavily-connected clumps. We would never be able to move a whole clump from one set to the other because moving the first cell would give negative gain, even though moving a few at once gives a positive gain. To avoid this *local minimum* of the cut count, we compute the gains for exchanging multiple pairs of cells and exchange the number of cells that improves the cut count the most.

```
gmax := 0;  gainSum(0) := 0;
do begin
  A' := A;  B' := B;
  for i := 1 upto n/2 do begin
    compute D for all cells
    find a and b in A' and B' such that gain is maximized
    exchange a and b in A', B'
    ignore a and b for the rest of this pass
    gainSum(i) := gainSum(i-1)+gain
    if gainSum(i)>gmax then begin
      gmax := gainSum(i);  k := i;
    end;
  end;
  if gmax<=0 then done;    # no improvement
  exchange a(1), ..., a(k) for b(1), ..., b(k) in A, B
end;
```

After min-cut finds a good initial placement, we run a placement improvement function to find a better final placement. We measure the quality of a placement with a *cost function*. The most common cost function is the total length of all wires on the chip. We find positions of the connectors on the cells and approximate the length of wire for each net by half the perimeter of the bounding box of the connectors on the net. We use total net length as our cost function because it is related to the total chip size (shorter wires will probably lead to less congested channels) and to chip performance (shorter wires overall will probably lead to shorter wires along critical timing paths). We can improve the quality of the cost function by adding terms for channel density and wire length along critical paths, but such additions slow down the calculation of the cost function. Because we calculate the cost function very often, we prefer a simple function that is cheap to calculate and maintain.

The most common placement improvement method is *pairwise interchange*. We choose two cells in the chip and consider the effect of exchanging them. If the overall cost function improves, we accept the change, otherwise we reject it and continue. One pass through pairwise interchange scans through all $n(n-1)$ pairs of cells taking every cost improvement. We may repeat this scanning until there is no more improvement, but since n may be large, we usually impose some maximum number of passes

Performance of this algorithm degrades significantly with large n. We can reduce this problem by limiting the exchanges to cells that are already within some preset distance of each other. Cells won't make large jumps, but we assume that because of the min-cut partitioning, they are already near their optimal final positions.

Because it is a greedy algorithm, pairwise interchange produces a result that may be only locally optimal. Just like with the min-cut partition, we may find that a specific exchange may worsen the cost function, but several may improve it. The most common techniques used to "climb out" of a local minimum are *monte carlo* randomization techniques. In *simulated annealing* (Kirkpatrick et al. 1983), a popular monte carlo technique, we accept all exchanges that improve the cost function plus some exchanges that degrade the cost function. In an algorithm patterned after the motion of molecules in thermal annealing, we accept an exchange if it increases in the cost function by Δc with probability $e^{-\Delta c/kT}$, where k is Boltzmann's constant and T is the temperature. In simulated annealing, we combine k and T into one factor for the temperature. As pairwise interchange progresses, we decrease the temperature slowly, accepting fewer and fewer degrading exchanges. Finally, the temperature goes to zero and we have the *greedy* case we already considered.

Although simulated annealing can produce a very good result, it is expensive in computer time and the quality of the result is dependent on the rate of cooling, which is dependent on the data. It also takes a large amount of computer time compared to other, less-random algorithms. Nahar et al. (1985) showed that a system that does less randomization will produce a better result in the same amount of time. Therefore, production systems do as little randomization as possible, usually optimizing with greedy interchange, saving the solution and randomizing when there is no more improvement to be gained.

Given sufficient time, monte carlo techniques can produce a placement with a very low cost. However, the cost function is only indirectly related to the size and speed of the chip. Extreme optimization of wire length produces channels that are terribly congested, perhaps destroying the routability of the chip. It is usually more advantageous to spend computer time calculating a better cost function than it is to spend the time optimizing a poor one.

We have treated all cells as if they are all the same size. In fact, cells are variable-sized, so exchanging them one-for-one does not preserve the size the sets during the min-cut placement or the length of the rows during placement improvement. We address these problems by rejecting exchanges in the min-cut that change the overall sizes of sets by more than some preset amount. To reduce row length differences during placement improvement, we can include the row length variation in the cost function.

There are several improvements we can make to the placement algorithm. First, we can mark nets that are critical by giving them a *weight*. A net with weight of 10 would count as heavily as ten nets in the min-cut and placement improvement, so we would be more likely to optimize that net than some other. We can improve the initial placement by taking into account the relative positions of the sets of cells when we compute the subdivisions (Dunlop and Kernighan 1985 and Hartoog 1986). We can also take into account mirroring cells in the placement improvement.

Global Route

A *global net* is a net that has connectors in more than one channel. The general problem of global routing is to determine through which channels each global net should pass. In the case of a standard cell floorplan, the problem is simplified because we know the channel structure in advance.

Most standard cell systems allow a global net to cross a row of cells by inserting a *feedthrough*, a gap through the cell row where a single net may cross into the next channel. In a standard cell system, the global router's job is to assign nets to feedthroughs to allow every global net to enter all channels in which it has a connector. A feedthrough may be used by only one net, but usually the global router has the capability to add feedthroughs if it needs them. As we have defined our standard cells, the signals connect on the top and the bottom, so each net has a free feedthrough in every row where it has a cell.

In practice, a greedy algorithm that takes any feedthrough within the bounding box of the net works adequately, but a good global router may attempt to optimize the assignment of feedthroughs to avoid channel congestion and hence ease the job of the channel router.

Channel Routing

We have a rectangular channel with rows of connectors on top and bottom. Each connector is part of a net. The channel router connects all connectors in all nets. We have two wiring layers. In this section, we investigate one method of addressing channel routing in which we route a net horizontally along the channel in one layer and connect all connectors on that net with vertical segments on the other layer (Yoshimura and Kuh 1982).

Layout Generators

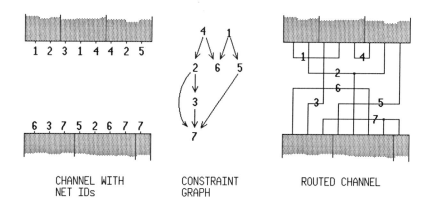

CHANNEL WITH NET IDs　　CONSTRAINT GRAPH　　ROUTED CHANNEL

The first step in channel routing is *track assignment* during which we assign the nets to tracks. First, we determine the *span* of all nets. The span of a net is the interval from the leftmost connector to the rightmost connector. We can assign several nets to the same track if their spans do not overlap.

We cannot choose the ordering of tracks in the channel arbitrarily. For each pair of connectors, one on top of the channel and one on the bottom, we must order their nets so that the net of the top connector is above the net of the bottom connector. If we do not, we short the two nets together when we make the vertical segments to the connectors. We can express this partial ordering as a graph with nets as nodes and constraints as arcs. We make a constraint from net 1 to net 2 if net 1 has a connector that is above a connector on net 2. This graph is the *channel constraint graph*. We can pack nets into tracks as long as we maintain the ordering in the constraint graph. When the track assignment is done, we route the wires in their tracks and connect to them vertically to their connectors.

If the constraint graph has a cycle, there is no legal track assignment. There are two methods for addressing this problem. We can introduce a *dogleg*, where a single net changes tracks part way down the channel, breaking the vertical cycle. Sangiovanni-Vincentelli et al. (1984) recommended that we break the cycle by ignoring one of the constraints, route as much of the channel as we can, and try to finish the route with a *maze router*. A maze router searches for any possible legal connection on any layers, so if it is possible to finish the route, it will.

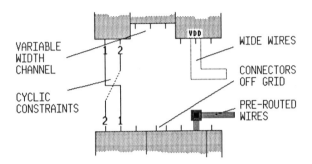

There are many other complications to channel routing. The channels may not be rectangular, so we may have some partially-blocked tracks to use. We may want to route wires that have with different widths, so the tracks may not be evenly spaced. The connectors on the top and bottom of the channel may not be on a grid, so we must move their wires onto a grid or route without a grid. There may be wires in the channel that a user already routed that we must avoid.

There are many methods for improving the quality of the route. If track assignment gives a result that is much greater than the density of the channel, doglegging may be used to break long spans that are holding the channel apart. If we have more than two wiring layers, we can route tracks on top of one another. Finally, we assumed that the tracks were separated enough that we can put contacts on adjacent tracks and not have a design rule violation. If we have no situation where two contacts are adjacent, we can *compact* the channel, spacing the two tracks more closely.

Other approaches to channel routing, such as the Rivest and Fiduccia's greedy approach (1982) and Burstein and Pelavin's hierarchical approach (1983), address these problems with some success. However, these algorithms have other disadvantages, such as the creation of large numbers of contact vias which make them less attractive.

Placement and Routing of Gate Arrays and Arbitrary Blocks

Gate arrays are very similar to standard cells. They break nicely into cell rows with channels between them. However, the channels are fixed in size. In the standard cell system, we can add as many tracks between rows as we need to make all the wires. In the gate array, if we exceed the predetermined number in the channel, the route fails. Gate array routers must have a good global router that takes into account channel congestion.

A channel router for a gate array must have a method for attempting to complete routes in channels that are too full.

Arbitrary block chips have no simple channel definition, because the cells may not span the whole chip, so we may find channels that have connections on more than two sides. Such a channel cannot be routed with the channel router we have described. A common method to address this problem is to adopt a *slicing floorplan* where the set of cells can be cut by a horizontal or vertical slice. By recursively dividing the cell, we can fully slice the chip into one-cell blocks. We route the cells in the reverse order of the slices, so each route is a channel route.

Arbitrary block placement is complicated by the irregularity of the blocks and by the further constraint that we must build a rectangular chip as well as try to minimize wiring. Global routing searches a graph representation of the channel connections to find the shortest connections for global nets through the topologically-complex arrangement of channels. Finally, in an arbitrary block system, we must route VDD and VSS to all cells. Their wires must be wide enough to carry the current, and if they cross, they may require a huge contact structure. Power routers can be very complex.

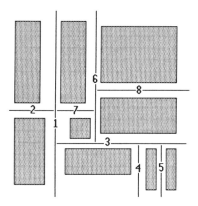

Exercises

Programming Problems

1. Make a 2-input NAND gate with variable-sized transistor width using the instance overlay and language based method.

2. We implemented the PLA with the instance method with overlays. Write code for the PLA core using a pure instance method.

3. Write a translator from boolean logic equations like those on page 154 to a PLA code file. What restrictions must you put on the equations?

4. Modify the PLA code file to take arguments to optionally include higher-power output drivers. Modify the code of your PLA generator to size the input buffers depending on the number of transistors they have to drive.

5. Write a heterogeneous array generator. A heterogeneous array generator builds a rectangular array of identically-sized cell. The specification is a code file similar to the PLA file. The array generator code file has three parts: the array dimensions, definitions for a mapping from characters to cell names, and the body in which the characters represent instances of cells in the array.

```
# 3 BY 4 ARRAY
3 4

# 3 DIFFERENT CELLS
3
A DIAG
B OTHER
C BOT

# THE ARRAY
A B B
B A B
B B A
C C C
```

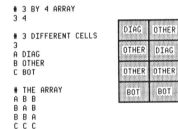

Modify your system to build arrays of differently-sized cells as long as all cells in a column have the same width and all cells in a row have the same height. Can you build a PLA with the array builder? A datapath?

6. A *river router* connects two horizontal rows of connectors on a single layer. It outputs wires on the layer that meets all design rule constraints and connects the two lists of positions in order. There is no restriction on the vertical size of the resulting route. Write a river router and incorporate it into the layout language.

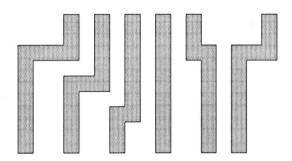

Questions

1. The relative sizes of design rules change with respect to one another. How would you make cells in the layout language that are insensitive to design rule changes? Explain your solution for both the programming style and the instance style.

2. Modify the PLA code file BNF to allow specifying the number of feedbacks and the sizes of output drivers.

3. How would you build a PLA that can have an odd number of product terms and outputs?

4. A criticism of datapath compilers is that the cells leave space for all their busses even if a particular application doesn't need them. How do you make a datapath compiler that doesn't leave space for busses when it doesn't need them?

5. Automatic placement and routing systems do not always complete the routing. What sort of tool would you provide for your users to assist in completing the routes?

References

PLA Construction and Optimization

R.K. Brayton, G.D. Hachtel, C.T. McMullen, A.L. Sangiovanni-Vincentelli, *Logic Minimization Algorithms for VLSI Synthesis*, Kluwer Academic Publishers, 1984.

G.D. Hachtel, A.R. Newton and A.L. Sangiovanni-Vincentelli, "Techniques for Programmable Logic Array Folding", *Proceedings of the 19th Design Automation Conference*, 1982.

Silicon Compilation

R. Ayres, *Silicon Compilation and the Art of Automatic Microchip Design*, Prentice Hall, 1983.

D.L. Johannsen, *Silicon Compilation*, Ph.D. Thesis, California Institute of Technology, 1981.

J. Rowson, B. Walker and S. Dholakia, "A Datapath Compiler for Standard Cells and Gate Arrays", *Proceedings of the 1987 Custom Integrated Circuits Conference*.

Placement and Routing

M.A. Breuer, "Min-Cut Placement", *Journal of Design Automation and Fault Tolerant Computing*, v1, no 4, October, 1977.

M. Burstein and R. Pelavin, "Hierarchical Channel Router", *Proceedings of the 20th Design Automation Conference*, 1983.

A.E. Dunlop and B.W. Kernighan, "A Procedure for Placement of Standard Cell VLSI Circuits", *IEEE Transactions on Computer Aided Design of Integrated Circuits and Systems*, vol. CAD-4, no. 1, January 1985.

M.R. Hartoog, "Analysis of Placement Procedures for VLSI Standard Cell Layout", *Proceedings of the 23rd Design Automation Conference*, 1986.

T.C. Hu and Ernest S. Kuh, ed., *VLSI Circuit Layout: Theory and Design*, IEEE Press, 1985.

B.W. Kernighan and S. Lin, "An Efficient Heuristic Procedure for Partitioning Graphs", *Bell System Technical Journal*, vol. 49, no. 2, February, 1970.

S. Kirkpatrick, C.D. Gelatt, Jr., M.P. Vecchi, "Optimization by Simulated Annealing", *Science*, vol. 220, No 4598, 13 May 1983.

S. Nahar, S. Sahni, E. Shragowitz, "Experiments with Simulated Annealing", Proceedings of the 22nd Design Automation Conference, 1985.

R.L. Rivest and C.M. Fiduccia, "A Greedy Channel Router", Proceedings of the 19th Design Automation Conference, 1982.

A. Sangiovanni-Vincentelli, M. Santomauro and J. Reed, "A New Gridless Channel Router: Yet Another Channel Router the Second (YACR-II)", *Proceedings of the International Conference on Computer-Aided Design*, 1984.

J. Soukup, "Circuit Layout", *Proceedings of the IEEE*, v 69, Oct 1981.

T. Yoshimura and E.S. Kuh, "Efficient Algorithms for Channel Routing", *IEEE Transactions on Computer Aided Design of Integrated Circuits and Systems*, vol. CAD-1, no. 1, January 1982.

General

D. Edgington, B. Walker, S. Nance, C. Starr, S. Dholakia and M. Kliment, "CMOS Cell-Layout Compilers for IC Design", *Proceedings of the IEEE Custom Integrated Circuits Conference*, 1984.

A. Martínez and S. Nance "Methodology for Compiler Generated Silicon Structures", *Proceedings of the 21st Design Automation Conference*, 1984.

R.N. Mayo and J.K. Ousterhout, "Pictures with Parentheses: Combining Graphics and Procedures in a VLSI Layout Tool", *Proceedings of the 20th Design Automation Conference*, 1983.

J.A. Newkirk and R. Mathews, *The VLSI Designer's Library*, Addison-Wesley, 1983.

J.A. Rowson, *Understanding Hierarchical Design*, Ph.D. Thesis, California Institute of Technology, 1980.

S. Trimberger "Combining Graphics and a Layout Language in a Single Interactive System", *Proceedings of the 18th Design Automation Conference*, 1981.

J.D. Ullman, *Computational Aspects of VLSI*, Computer Science Press, 1984.

CHAPTER 8

LAYOUT ANALYSIS

So far, we have concentrated on tools for synthesis -- tools that create layout. We now consider analysis tools, tools that check that the layout is correct. This chapter describes two tools for verification of layouts: a design rule checker (DRC) and a circuit extractor. A design rule checker checks that the widths, spacings and overlaps of features in the layout meet some minimum criteria, the *design rules*. A circuit extractor reads a layout and writes a netlist of transistors, capacitors and resistors.

Although these two tools have rather different functions, their implementations share some important fundamental algorithms, so we discuss both in this chapter. Both DRC and circuit extraction accept as input mask layout data and both tools face the same problem constraints due to the mass of data. In addition, design rule checking and circuit extraction are both used at the same point in the design cycle, when the layout is nearing completion and the user must check that his work is correct.

Overview and Background

A major difficulty with these checking operations is that they cannot be simply separated by hierarchical decomposition of the design. The job of a design rule checker is to examine the sizes and separations of shapes on layers and between different layers to determine whether or not they meet certain minima -- the *design rules*. Two edges are too close even if they are inside different cells. One cell placed twice must be understood as contributing to two separate circuits during circuit extraction. Therefore, layout analysis tools work on a "flat" representation of the layout, with all hierarchy removed. This flat representation comprises an immense amount of data, so the algorithms must be very space and time efficient.

We wish to be *conservative* in our analysis. The penalty for failing to report an error is much greater than the penalty for reporting an error when there is none. Therefore, when we are faced with a decision in algorithm design, we will accept a solution that will catch real errors, but possibly report *false errors* as well. False errors are not without cost. Like the boy who cried "Wolf!", our software must not report so many false alarms that users ignore real errors. We attempt to reduce false errors, but not to the point where we might not report a real error.

The software we use for layout analysis consists of a set of two-dimensional manipulations plus a set of checks on the result of those manipulations. We start this chapter with a discussion of a simple checking algorithm and investigate performance problems with it. We then discuss edge-based techniques for design rule checking and circuit extraction.

Design Rules

Design rules arise from limitations in the integrated circuit manufacturing process. These limitations are primarily due to the limits of the resolution and precision of alignment of photolithography equipment. They also arise from consideration of the physical nature of masks that define the patterns on the circuit in an attempt to reduce defects during mask making.

The following is a graphical representation of the scalable design rules for the MOSIS single-level-metal CMOS process taken from Mukherjee (1986). The dimensions are given in units of lambda. The design rules specify:
- Minimum width of features on all layers.
- Spacing between unrelated features on layers.
- Spacing between polysilicon and diffusion.
- Spacing between contact and transistor (polysilicon + diffusion).
- Minimum extension of polysilicon and diffusion past transistor gate active area.
- Minimum overlap of metal, polysilicon and diffusion around contact cuts.

Checks are specified between drawn layers and between *derived* layers, layers made up of combinations of the drawn layers. Transistors, for example, are areas where both polysilicon and diffusion are present. Non-transistor polysilicon (polysilicon minus diffusion) is another derived layer. Our analysis software must perform boolean operations on the masks: OR, AND and NOT.

Often, there are pairs of rules, only one of which need be obeyed. For example, polysilicon and diffusion must be separated by 1λ except when they make a transistor, in which case they must overlap by at least 2λ.

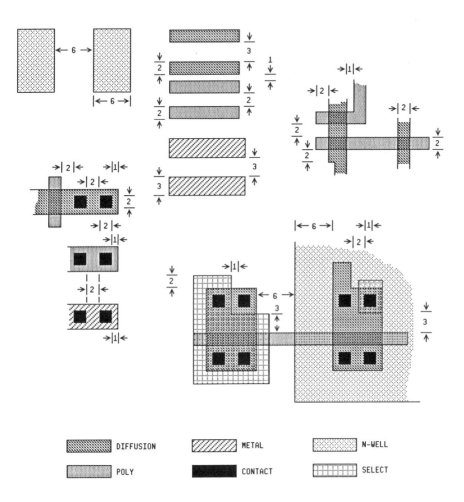

Object-Based DRC

We can implement a very simple checking program by scanning the lists of graphical objects in our data structure and checking each object independently for minimum width and pairs of objects for spacing.

Width Checks

For each rectangle, we ensure that both dimensions are greater than the minimum. However, that check is incorrect for rectangles that touch. On the left in the following figure, the check would report a width violation for each of the two rectangles when, in fact, together they make a legal figure. In the second case, the check would report no error, when there is a minimum width violation. Obviously, a much better width check is required.

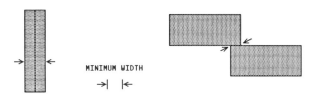

Spacing Checks

We can implement an object-based spacing check by comparing all objects against all other objects and verifying that they do not violate the minimum spacing rules between them. However, we would flag an error for the following case where there are two rectangles too close to one another. Since the two rectangles are connected to one another, there may be no violation.

Merging Connected Rectangles

We can solve the problems with connected rectangles in the spacing check and some of the problems of the width check by first making a pass over the cell merging connected rectangles into large polygons. However, we are then faced with the task of checking width rules in a polygon.

Time and Space Requirements

As we already mentioned, we must be able to perform boolean operations on full mask layers. These operations include merging shapes into large polygons, ANDing masks, ORing masks and finding the difference between masks. Because of the large amount of data, these operations must be efficient.

Our naive spacing check requires that we compare each object against all others. The complexity of this algorithm is $O(n^2)$. In addition, we must scan through all edges on a layer many times. This requires reading that data from a disk file many times. Our program would be slowed down significantly by I/O. Obviously, sorting would help, but the order of complexity would still be $O(n^2)$.

Quad trees described in Chapter 5 could reduce the number of comparisons to $O(n\log n)$, but quad trees require significant overhead in data structures and random access to the data structure. Since the data will probably not fit into the physical address space of our computer, we must work from a disk file. Random file access is very slow. Scott and Ousterhout (1986) used corner stitched tiles as a form from which to perform design rule checking, but corner stitched tiles do not handle arbitrary angles well.

The algorithms described in the remainder of this chapter use a simple structure to avoid the quad tree overhead and random access. They are able to handle arbitrary angles in the layout. Some searching and sorting of large files is required, but the structures are well suited to the operations we do.

Edge-Based Layout Operations

Design rules specify limitations on edges of objects, not the objects themselves. A key idea is to break polygons and rectangles into their constituent edges and work on the edges (Baird 1977). At first, this seems counterproductive, since there are many more edges than there are objects, but edges are simpler. Each one requires only a few bytes and we will see that we need not represent all edges.

The first step of the DRC program is to read a layout file, flattening the hierarchy by expanding instances and converting polygons and boxes to edges. This hierarchical expansion uses the same kind of recursive algorithm we used in plotting. We sort the edges and write them to *edge files* since, in general, there will be many more edges than will fit in memory. We will make one edge file per layer in the layout. Subsequent passes process the data left to right over the chip. We read edge files, perform some checking operations and write new edge files. Finally, we translate the edges back into polygons for error reporting and display.

Edges

An *edge* consists of a start point and an end point. We order edges so the start point X is always less than or equal to the end point X. The edge has a bit indicating which side of the edge is the filled area. This bit is called a direction, since we can consider edges going clockwise around a filled area. A set of edges going counterclockwise around a region represents a hole, an empty area. The direction bit indicates that the filled area is above or to the left of the edge. Depending on the operation we are performing, the edge may have other information, including a polygon or node identifier.

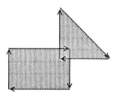

Removing Vertical Edges

Lauther (1981) pointed out that we can safely omit the vertical edges from the edge files without losing any information. To convert the edges back to layout, we use the other edges and their direction bits to infer the presence of the vertical edges. In the following figure, the filled area is above edge a and below edge b. We know that there must be a vertical edge bordering a filled area, so we can infer the presence of vertical edges from the remaining edges in the file. At each X location where non-vertical edges begin and end, there is an implied vertical edge between every adjacent pair of edges that have layout between them. Since nearly all layout is orthogonal, nearly half the edges are vertical. This observation eliminates approximately half of the edges from our data files.

We eliminate vertical edges because we process edges left to right. If we had intended to process edges bottom to top, we could have used the same argument to eliminate horizontal edges instead of vertical edges.

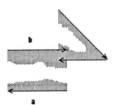

Sorting Edges

We make one list for each layer and sort the edges by increasing minimum X coordinate. When we do our checks, we will traverse this list, checking the chip from left to right. At any one time, we examine a specific X coordinate, so we only need to keep in memory the edges that interact at that coordinate. Statistically, we have only $\sqrt{n}$ objects to compare at each coordinate, reducing the overall complexity to $O(n^{1.5})$.

To improve the performance of further algorithms, we sort first by X, then by Y, then by increasing slope of the edge. Referring to the following figure, the order of the edges is d, b, g, e. Edges a, c and f are vertical edges and are removed from the scan.

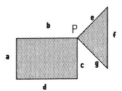

The Fundamental Algorithm

We pass a vertical *scan line* across the chip from left to right, examining only those edges that cross the scan line. At each scan line position, we build a partial solution in the form of edges and output the edges when the scan line has passed them completely. We have three main data structures: the sorted list of edges from the edge file, a priority queue of edges that we create and a list of all edges that are *active*, that is, edges that cross the current X position, the scan line. By keeping only the edges crossing the scan line, we limit the number of edges through which we search to do our operations.

The scan line moves from left to right stopping at beginnings and ends of edges. For each new scan line position, we add to the active list all edges that now touch the scan line and we remove all those that now lie completely to the left of the line. During the processing of one scan line, we determine the next scan position.

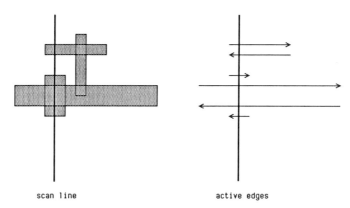

scan line   active edges

We are finished with an edge when the scan line has passed its right end, so long edges remain in the active list for a long time. The edge operation does not leave the edges sorted by their left coordinate, but by their right coordinate. We must re-sort the edge file before performing further operations.

```
currentX := -infinity;
nextX := infinity;
while currentX<infinity do begin
  if nextEdge.x <= nextX then begin
    addEdge(nextEdge,activeList);
    currentX := nextEdge.x;
  end else currentX := nextX;
  nextX := infinity;
  for all edges (e) in activeList do begin
    if e.highx<currentX then removeEdge(e,activeList)
    else begin
      nextX := nextX min e.highx;
      processEdge(e);
    end;
  end;
end;
```

Layout Analysis

Handling Intersecting Edges

We keep the edges in the active list sorted by increasing Y. In procedure addEdge, we sort the new edge into the active list. If the new edge intersects an existing edge adjacent to it in the active list, we calculate the point of intersection and break both edges at the intersection point. The left sides of both edges remain in the active list, the right sides are both wholly to the right of the scan line, because the edge we just added has its lower X end first. We put these new edges into a priority queue and merge them with the stream of input edges as we need new edges.

When we remove an edge with removeEdge, we check the newly-adjacent edges for intersection, breaking the old edges and adding the new ones to the sorted list of edges. The list of active edges remains sorted. If we did not break edges at intersections, crossing edges would un-sort the active list and we would be forced to re-sort it at every new scan line location.

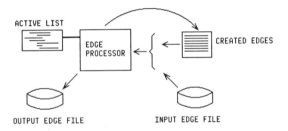

Polygon Merging

We now examine the basic operation of the edge-based algorithms by considering the polygon merging task. The algorithm is relevant not only to polygon merging but also to scan conversion and interior line removal, as discussed in Chapter 4.

The goal of polygon merging is to remove redundant internal edges in the layout, as shown in the following figure. We output only the outside edges of the layout, those in the following figure on the right. A single edge in the input file may become more than one edge in the output.

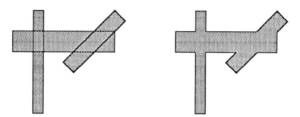

We distinguish area "inside" from area "outside" by the *wrap number*. To find the wrap number of a region, we start with any point in the region. The wrap number is the number of clockwise edges crossed by an imaginary line from that point to ∞ in any direction, minus the number of counterclockwise edges crossed. Since we know that ∞ is outside all areas, a region is "inside" if the wrap number is greater than zero. On the left of the preceeding figure, the open areas have wrap number 0, the regions that are part of only one rectangle have wrap number 1 and the intersection areas have wrap number 2. In the merged figure, all regions with wrap numbers greater than 0 now have wrap number 1.

As we scan through the edges in the active list from bottom to top, we keep an integer `wrapNumber` that tells us the current wrap number for the region we are in along the scan line. `wrapNumber` is zero at the start of scan line processing. We change `wrapNumber` every time we find an edge, incrementing or decrementing depending on the direction of the edge.

When we change `wrapNumber` to zero or from zero, we know that the edge that caused the change is a border between filled area and empty area, so we mark the edge as being a border. When we remove an edge from the active list, we check the border flag on the edge. If the edge is a border, then we output it. If it is not a border, we discard it because it is an interior edge in the layout.

We also check an edge when we change it from *border* to non-*border*. When that happens, we output the part of the edge from the start of the edge to the current location. Conceptually, we split the edge into two pieces: the part from the left end to the current scan line position that is a *border* and the part from the current position to the right end, which is non-*border*. We output the first part of the edge and remove it from the active list, since the scan line has now passed that location. We keep the later part of the edge in the active list.

When an edge changes from non-*border* to *border*, we remove the non-border portion of the edge, since that part of the edge will not appear in the output file. As in the case above, we can think of it as splitting the edge into a border and non-border part. Since the non-border part is entirely to the left of the scan line, we remove it from the active list and discard it. The active portion starts at the scan line and extends to the right. For efficiency sake, we need not create a new edge record, but only determine the new Y coordinate for the current scan line X coordinate for the edge and replace the start point of the edge with the new X, Y values.

Let us examine polygon merging in detail. The unmerged rectangles from the preceeding figure are transformed into the following edges on the left. We want to output only the edges on the right.

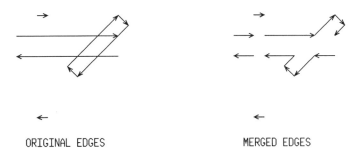

ORIGINAL EDGES   MERGED EDGES

Let us examine the progress of the polygon merge as the scan line moves from left to right across the layout. We refer to the following annotated figure, where each edge has a letter for its name and the stopping points of the scan line are numbered. Edges in the active list are bold.

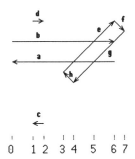

When we start, the active list is empty. And, of course, no edges are *border* edges. We read the first edge, edge **a** from the edge file, and move the scan line to location 0.

`active list: empty`

At Location 0

We add edges **a** and **b** to the active list and check the filled area. On each scan line, `wrapNumber` starts at zero at $y = -\infty$. Scanning through the active list, we find edge **a** running right to left, so the area above it is filled. Since `wrapNumber` is zero, we set edge **a** *border* and set `wrapNumber` to 1. Edge **b** runs left to right, indicating the end of a filled area. We decrement `wrapNumber` and, since it is zero again, we set **b** *border*.

active list: a b

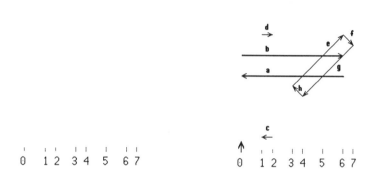

At Location 1

We add edges **c** and **d** to the active list. Scanning through the active edges, we first find edge **c**, increment `wrapNumber` and set **c** *border*. Next, we find edge **a** and increment `wrapNumber` to 2. Since `wrapNumber` was nonzero when we found edge **a**, edge **a** is no longer a border. We output the part of **a** from the left end to the current X position and change edge **a** to non *border*. We handle edge **b** similarly, outputting the leading segment and changing it to non *border*. At edge **b**, we decrement the wrapNumber back to 1. Finally, we reach edge **d**. We decrement `wrapNumber` again and find it reaches 0, so edge **d** is a *border*.

active list: c a b d

Layout Analysis   195

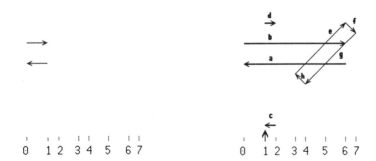

At Location 2

We do not add any edge at location 2. We determined that we must run the scan line through this point because edges **c** and **d** end at this location. As we scan the active list, edge **c** ends. Since it is *border*, we write it to the output file, and `wrapNumber` is still 0. When we find edge **a** we see that it must be a border now. Since it was not previously a border, we truncate the edge to the current X value, cutting off the part that we will never see again. We call the new part of the edge **a'** and set it to be a *border*. Edge **b** is handled similarly to edge **a**. Edge **d** is handled similarly to edge **c**.

`active list: a' b'`

At Location 3

The scan line moves to location 3 where we add two diagonal edges, **h** and **e**. Edge **h** has a smaller slope (-1), so it is in the edge file before edge **e**. When we add **e**, we notice intersections with its adjacent edge, **a'**. We made this check for all edges when we added them, but edge **e** is the first edge that has intersections.

We split the intersecting edges in two, as shown in the following figure. The bold edges are in the active list of edges, the remaining edges are still awaiting input. We did not attempt to intersect edge e' with edge b' because they were not adjacent in the active list. All edges in the active list are border edges.

activelist: h e a' b'

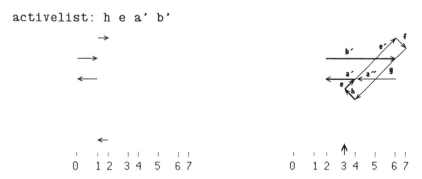

At Location 4

Edges **h**, **e** and **a'** end at location 4. They are all border edges, so they are all written to the output. Edges **g**, **e'** and **a"** are added to the active list. Edges **g** and **a"** intersect, so they are split and **g'** and **a'''** are added to the list of edges waiting to be added to the active list. When we add edge **e'**, we find that it intersects edge **b'**, so we split them also.

activelist: g a'' e' b'

At Location 5

All edges in the active list end. Edges **g** and **b'** are the only border edges, so they are written. New edges are **a'''**, **g'**, **b"** and **e"**.

activelist: a''' g' b'' e''

At Location 6

Edge **a'''** ends and is a border, so we write it. Edge **b"** ends, but is not a border. Edge **g"** is a new edge and a border. Edge **e"** ends and is written, and edge **f** is a new border edge.

activeList: g'' f

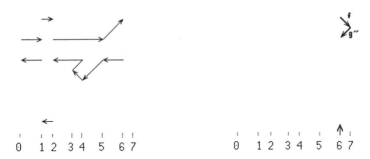

At Location 7

Edges **g"** and **f** end and are border edges, so they are written. There are no more edges, the active list is empty, and the merge ends.

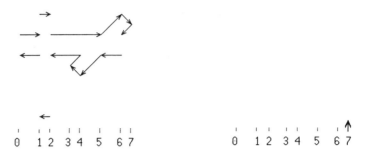

Arbitrary Boolean Operations On Layout Layers

The merging operation is the simplest operation we perform on edge files. We can use the same basic algorithm to produce more complex operations on edges. During polygon merge, we made an edge a *border* when the wrap number changed to or from zero. We can use different conditions on the wrap number to produce different functions on the output. For example, we can make an edge a border if the wrap number is greater than 1. This will identify all overlap areas. We can use this to identify overlaps between two sets of previously-merged edges.

We used the wrap number and border to merge all polygons. We made the assumption that all polygons were on the same layer. If we merge polygons from two different layers together, the result is the logical OR, all area covered by all polygons. If we set the border when the wrap number for either layer changes to or from zero, we also get the OR. If we set the border when the wrap number for either layer changes to or from zero and the other layer is greater than zero, then the result is the AND, the overlap of the two layers. To negate a layer, we add edges from $-\infty$ to $+\infty$ at $y = -\infty$ and $y = +\infty$, and take the inverse of the direction of every edge in the file.

We keep separate wrap numbers for each layer and write functions using the wrap numbers to produce complex boolean functions of the input layers. We need only compute the function of the presence or absence of the layers themselves along the scan line. We can merge and perform several boolean operations simultaneously, keeping separate *borders* for each of the functions we compute. This saves additional passes through the data.

Resizing Layout

Another common operation on layout is resizing the layout. Resizing, also called *bloating* and *shrinking*, is the operation of enlarging and shrinking all layout by a fixed distance. For example, we may wish to enlarge all layout by 1λ. This is not the same as scaling, since the filled area becomes larger, but the empty area shrinks.

ORIGINAL   1λ SHRINK   1λ BLOAT

One way to resize layout is to enlarge all polygons, moving edges by the required distance. A rectangle with coordinates -1 1 15 10 becomes -2 0 16 11. Polygons are trickier, since we must move all edges perpendicular to their direction and re-compute the edge intersection points.

Since we are dealing with edges already, we can perform the sizing operations on the edges. Resizing is more difficult with edges than with objects because we are moving the vertex between two edges. If we have objects, we have the two edges. If we work with edge files, we must find pairs of edges that meet at a vertex before we determine the new position of the vertex.

When we add a new edge, we look through the active list for the left *mate*, the edge that forms the vertex at the left end. If there is no mate, the edge connecting to the new edge is vertical, so we find the new coordinate assuming the presence of the vertical edge that is indicated by the wrap number crossing the new edge. We calculate the new intersection point based on the direction and slope of the two edges. When the scan line passes the right end of an edge, we perform a similar search for the right mate. We write an edge when the scan line passes its original right end coordinate.

The output of resizing is not sorted or merged, so we re-sort and re-merge the file before performing further operations.

Problems With Acute Angles

When we expand a set of edges, we would like the new figure to include all the area within the expansion distance of the old figure. If we wish to be precise about this, it means that the corners of a rectangle grow into circles, instead of the larger rectangle we made. Even though the corner grew as $\sqrt{2}$ times the expansion distance, this larger rectangle is acceptable since the difference is not great and we retain the advantage of working with rectangles.

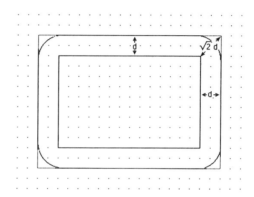

However, when we expand an acute angle, the distance of the new vertex from the old vertex can be arbitrarily large. A very acute angle results in a very distant point.

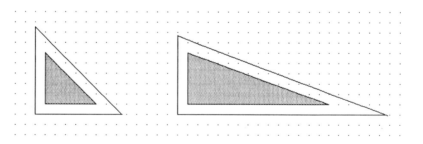

A very acute notch results in a spike the other direction, as shown in the following figure. We find the new vertices of the figure by moving the edges and finding the new intersection points. When we move a and b by 3 grids, we get the new vertex, ab'. After doing the same with edges b and c, we find that the new intersection point is well out to the right. The final enclosed area is the area shown shaded at the bottom right. The area inside the protruding triangle is wrapped twice in the figure, but polygon merging fixes that.

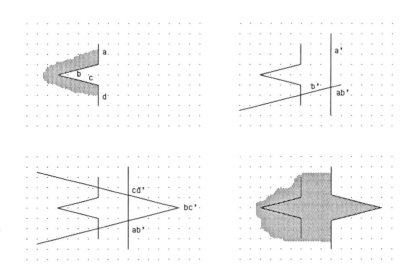

These spurious spikes lead to incorrect analysis and can be disastrous in chip layouts. There have been several techniques proposed to address the problem by recognizing acute angles and truncating the corners. These techniques add edges, potentially degrading the performance of the algorithm.

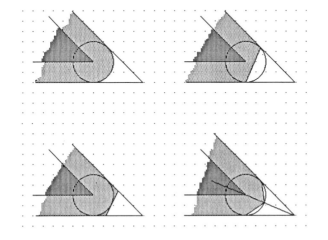

Using Bloat and Shrink to Perform a Design Rule Chec

As we have seen, design rules are expressed as minimum separations and minimum sizes of layouts. The merging, boolean and sizing operations we have seen are sufficient to perform these checks.

Spacing Checks

We start with fully expanded, merged layout. To determine if there are two features on the same layer too close, we expand the layout by half the design rule distance, then re-merge. This time, we output an edge only if the wrap number is 2 or more. These are the edges that were within the design rule distance in the original layout. If we expand the overlap polygons by slightly more than half the design rule distance and AND those edges with the original edges, we get markers for the offending edges.

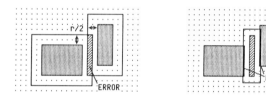

We can express our spacing check as code:

```
pointer(edgeFile) procedure checkLayer(
    pointer(edgeFile) edges;
    real designRule);
begin
    pointer(edgeFile) medges,temp,temp2;
    temp := merge(edges,wrapGT0);
    medges := sort(temp);
    temp := bloat(medges,designRule/2.0);
    temp2 := sort(temp);
    temp := merge(temp2,wrapGT1);
    if empty(temp) then return(nullpointer);   # no errors
    temp2 := sort(temp);
    temp := expand(temp2,designRule/2.0+epsilon);
    temp2 := sort(temp);
    return(andEdges(temp2,medges));
end;
```

To find design rule violations between edges on different layers, we expand both sets of edges by half the rule between them and AND the two layers. The resulting areas mark the design rule violations. We expand the areas and re-intersect them to find the edges in violation.

Minimum Feature Size Check

We check minimum feature size on a layer by shrinking by half the minimum size for the layer and finding all negative areas, areas with negative wrap count. The edges in the new rectangle at the right are counterclockwise, indicating that the original rectangle was smaller than twice the shrink size in one dimension.

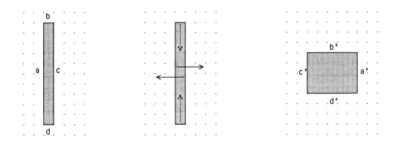

More Complex Rules

Rules that check contact overlap and spacing to transistors require that we find contact and transistor area. We find these areas by boolean operations on the mask layers we are given as input. For example, a transistor area is the area of diffusion AND polysilicon. We express the checking rule as minimum spacing between mask layers and the transistor layer that we create.

A More Efficient DRC

The resizing DRC works correctly most of the time, but it has several drawbacks. Most importantly, it can miss some errors. In addition, it requires several passes through the edge files with several sort and merge steps.

As we already discussed, bloating does not expand corners correctly, because corners move by more then the bloat distance. This leads to incorrect corner-to-corner checks. In addition, the minimum feature check misses very small rectangles because they invert twice to form a new positive area.

We can reduce the number of passes through the data by performing some of the checks simultaneously. Further, we need not expand layers if we check separation between the edges inside the processing code. So, we can perform all the checks of the DRC in one pass through the edges.

We avoid the problems of resizing by performing the width and spacing checks on the edges and corners we see in the edge file. We need not find touching pairs of edges, nor must we find new positions for the intersection points of edges.

We process all layers simultaneously with a separate scan line per layer. All edges that are closer to the scan line than the largest design rule involving that layer remain in the active list. We can view this as keeping a *horizon* for each layer. When the high end of an edge passes over the horizon, we can remove it from the list. In the polygon merge algorithm, the horizon was the same as the scan line, because edges had to touch to interact. In design rule checking, edges interact if they are closer than the design rule distance.

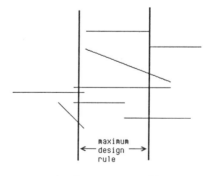

Layout Analysis 205

When we read an edge, we compare it with all edges in its own active list for proper width and with edges in the active lists of layers that interact with its layer for proper spacing. When we find a violation, we mark both edges and continue. We leave the edge in the active list until it can no longer violate any rules with new edges. That is the horizon line. When we remove an edge we it if it has been marked as a violation.

For each layer, we keep a pointer to the edge file, the scan line and the size of the largest design rule involving edges on the layer. The design rule size determines the horizon for that layer. In addition, we have a list of the edges in the active list on the layer. We sort the edges on the active list by maximum X, to make it easier to check for violations and to identify those edges that are past the horizon.

We express the actual design rule distances in two arrays. widths contains the minimum widths for each layer and spacings contains the minimum spacing between each pair of layers. When we read an edge, we check widths against edges on its layer and spacings against all layers.

```
e := nextEdge;
checkWidth(e,widths[i]);
for i := 1 upto numberOfLayers do begin
  edgeList := scanRec[i].activeList;
  for all edges (f) in edgeList do begin
    checkSeparation(e,f,spacings[e.layer,i]);
  end;
end;
```

Examining a New Edge

When adding an edge, we check it against all edges in all layers for which there is a design rule between the two layers. Since we know a violation must be evident at an endpoint, we perform four point-to-line distance checks to determine if the two edges are in violation.

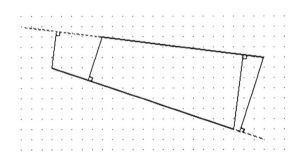

Because we are dealing with edges and not polygons, we determine from the edge directions whether to check widths or spacings between them. For parallel edges, the test is rather intuitive, depending on the orientation of the edges. For non-parallel edges, the answer is less so. We calculate the distance from the starting point of one edge to the other edge. For all non-orthogonal edges, we determine whether to use the width check or the spacing check by examining the relative orientations of the edges.

The details of this check involve some vector arithmetic. We find the point p where the perpendicular from the point we are checking intersects the other edge. We calculate δ, the distance along the perpendicular to the edge.

```
p = [x,y] + delta * [uy, -ux] = [xx, yy] + sigma * [uux,uuy]
```

where [ux,uy] is the unit vector along the first edge (the one we are checking the point on), so [uy, -ux] is the perpendicular to that edge. [uux, uuy] is the unit vector along the other edge. The unit vector along the edge directed from p1 to p2 is: $p2\text{-}p1/\sqrt{(x2\text{-}x1)^2+(y2\text{-}y1)^2}$.

The equation for calculating δ can be derived from the vector equation:

$\delta = ((x\text{-}xx)ux+(y\text{-}yy)uy/(uy\ uuy+ux\ uux)$

Delta is a signed value. If it is greater than zero, then the edges are oriented such that there is layout between them, and we do a width check. If delta is negative, then we do a spacing check.

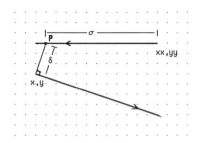

We must check separation of vertical edges, which are not represented in our edge files at all. If we do not, we risk missing the errors, as shown in the following figure. So, as we read edges, we infer the presence of vertical edges, build new edge records for them and include them in the active list to be checked with all other edges.

Layout Analysis

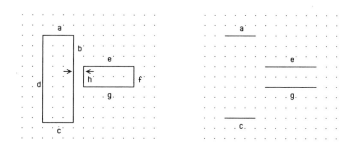

Derived Layers

Although we can derive transistor, transistor-overlap and contact areas as we check rules, our code is simpler if we separate the boolean operations on layers from the width and spacing checks. Our DRC becomes a two-pass operation, one to derive layers from the input layout and one to check the widths and spacings of all layers.

Enclosure Checks

The mechanism we have described does not perform enclosure checks. These are the checks that we must do to ensure that contact cuts, for example, are properly surrounded by metal. An enclosure check of one kind of material around another can be expressed as a minimum separation between one material and the absence of the other. Thus, we can reverse the directions of the edges of one layer to get the empty space and use a spacing check. Of course, there is no need to make another derived layer to represent the absence of a specific layer, we can include the overlap information with the spacing rule and invert the edges as we check the widths.

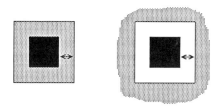

The Transistor Overlap Rule

The polysilicon gate of a transistor must overlap the transistor area by some minimum amount. Unlike the enclosure rule, the polysilicon must not fully surround the transistor area, it must extend only in two directions. To check the gate overlap, we require another kind of operation, which we call *crucify*. This operation extends all edges by some distance perpendicular to their directions. Existing endpoints remain and new edges are created. For example, a rectangle turns into a cross shape:

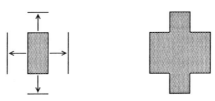

We use the crucify function to identify the area where polysilicon must be present, extending the overlap distance only in the direction of polysilicon. We do this in several steps:
1. Find all transistors by intersecting polysilicon and diffusion.
2. Crucify the transistors by a small amount, say $.1\lambda$.
3. AND the result with the polysilicon layer. This gives two small rectangles that identify which sides of the transistor have polysilicon present.
4. Crucify the layer from step 3 by the overlap amount.
5. Shrink the layout by the same small amount from step 1. This removes the segments perpendicular to the small sides of the polysilicon edges and it shrinks the extent in the polysilicon direction to exactly the overlap distance.

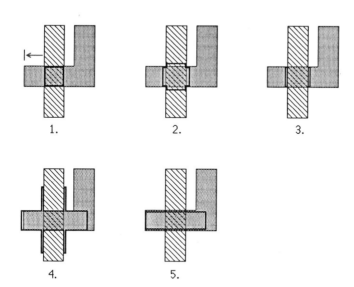

To check the polysilicon overlap, we check that polysilicon encloses the layer we just derived.

Disjunctive Rules

When polysilcon and diffusion cross, they make a transistor. To ensure that we get a transistor, we must ensure they cross by a minimum distance. If we do not want an accidental transistor, we must keep them separate by some other minimum distance.

If we check the separation of diffusion from poly, we will get a spurious separation violation message for every transistor. To solve this problem, we subtract from the polysilicon and diffusion layers the area around the transistor that would give the violation, then compare the remainder for design rule violations. For a rectangular transistor, the area we wish to remove is the transistor area crucified by the minimum spacing between polysilicon and diffusion.

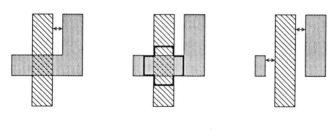

Original   Crucify Transistors   Subtract and Check

Connectivity

The width and spacing rules are different if the two features are part of the same electrical node than if they are disconnected. Usually, edges in the same node have no spacing rules. We would like to avoid the extra run time and false errors that arise because we use the disconnected-node rules for nodes that are in fact connected. We associate node numbers with edges in a process called *stamping*. Stamping node numbers on edges implies that we must derive the entire connectivity of the chip in order to check the design rules. Later in this chapter we describe a tool for extracting a netlist from the layout. We can use the extracted netlist for stamping.

Calculating Polygon Identifiers

We would like to avoid the performance problems due to a full circuit extraction while we are checking design rules. Fortunately, we can catch most of these problems by stamping each edge with an integer to uniquely identify its polygon rather than its full electrical node. We can identify polygons when we read the layout, so we can include that information with the edges. As we merge polygons, some polygons are absorbed, so we add the ability to generate unique polygon numbers to our scan line algorithm.

When we check edges, we check their polygon numbers to determine if we should use the connected-rules or the un-connected-rules. Using polygon numbers instead of true node numbers is conservative because it may generate more error messages, but the polygon number check is a good one because it does not generate many false errors. Still, if we are doing a design rule check and a circuit extraction at the same time, we could use the node numbers to reduce the number of false errors.

In this section we examine the method of Szymanski and Van Wyk (1983) for determining unique identifiers for polygons in the edge file. The same method can be applied to electrical node identification, both as a pre-processing step for design rule checking and as an integral part of circuit extraction, which we describe later in this chapter.

As we merge polygons, we keep *polygon records*, records that consist of an identifying integer and a set of the edges in the polygon. The opening and closing border edges are, of course, in the same polygon, as are all edges between them. These edges are all kept in the same polygon record. If the two border edges are already in different polygon records, we merge them. When two edges cross, the two edges must be in the same polygon, so we merge the polygon records. Similarly, when we add or remove an edge from the active list, we can discover sets of edges that we thought were in different polygons that are actually in the same polygon. That is the case in the following figure. As we move the scan line from left to right, we first see edges d and f in one polygon and edges a and b in another. When we write edges d and b, we write them with different polygon numbers. Later, when we edges e and c end, we know that the two polygons were really one polygon, so we merge them.

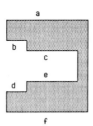

When we merge polygons, we throw out the polygon number from one polygon record and use the other. We are free to re-use the old polygon number as the number of another polygon. Problems arise when we want to change the polygon number for a polygon that has already had some edges written with the old number. We can write the new edges with the correct number, but edges that are already written cannot be changed. We handle this case by noting in the output edge file that we wish to *merge* polygon A into polygon B, meaning that wherever polygon A was used as the number of a polygon in the file, we should have used B. In addition, when a set of polygons becomes empty, we write the *end* of that the polygon with that number. The *end* record lets us re-use the polygon number without confusing the new use of the number with the old one. By re-using polygon numbers, we reduce the expected space usage of the second pass.

The preceeding figure would produce a file like this:

```
edge d, polygon 1
edge b, polygon 2
merge 2 into 1      we found end of e and c
edge e, polygon 1
edge c, polygon 1   all edges are in the same polygon now
edge f, polygon 1
edge a, polygon 1
end 1
```

The result of this pass through the edges is an edge file that includes extra *merge* and *end* records. We make a second pass through the file *backwards*. During the backward pass, when we encounter an *end N* record, we make a new entry to map polygons from number N to the next free polygon number. Subsequently, every polygon we find with number N, we will change to the new free polygon number. When we find a *merge A into B*, we set the final polygon number for A to be the current one for B. For every edge, we change the polygon number by indexing into the table of final numbers.

The code to do the second pass scan looks like this:

```
polygonCount := 0;
while not bof(input) do begin
  rec := reverseRead(input);
  if rec.type = endRecord then begin
    polygonCount := polygonCount + 1;
    finalNumbers[rec.polygonNumber] := polygonCount;
  end else if rec.type = mergeRecord then begin
    finalNumbers[rec.mergePolygonNumber] :=
      finalNumbers[rec.finalPolygonNumber];
  end else begin
    # rec must be an edge
    rec.edge.polygonNumber :=
      finalNumbers[rec.edge.polygonNumber];
  end;
end;
```

This algorithm is space efficient, using only $O(m)$ space in the final numbering, where m is the maximum number of polygons that are simultaneously cut by the scan line. This number is $O(\sqrt{n}\,)$ of the edges in the file. Had we not re-used node numbers in the first pass, the second pass would have required $O(n)$ space for the translation table.

Performance Optimization Considerations

General point-to-line distance calculations are time consuming. Since most of our edges are orthogonal, we can avoid the point-to-line calculations by treating horizontal and vertical edges as special cases. When we have two horizontal edges or two vertical edges to check, we can do a much simpler check. If the edges have a parallel overlap, we check the parallel separation of the edges. Otherwise, we check the distance between nearest endpoints.

Layout Analysis

We need never check perpendicular edges. They can be in violation at a single point, but the violation will be found between the edge that forms the bottom of the "L" and the connecting edge of the other edge.

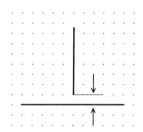

To facilitate these checks, we keep three lists of edges with each layer record, one for horizontal, one for vertical (which we create as we read the files) and one for diagonal edges. When we check edges, we perform simple checks between parallel edges, no checks at all between perpendicular edges and full checks between all other edges.

Reporting DRC Errors

The output of our DRC is an edge file listing all the edges that have been found to be in violation. However, we have not discussed how to present those edges to the user. One powerful way is to provide a translation back to layout files and to make a new layer for layout files that is the error layer. We translate each edge in the error file into a thin rectangle for easy viewing. A user reads the error file for a cell into the layout editor and overlays the errors on top of the layout for the cell. The layer on/off facility would let users optionally view the errors.

The problem with this method is that when a user views the error overlay, he knows which edges were in violation, not why they were in violation. We would like to mark each error edge with an identifier for the kind of check that was violated by the edge. When translating to layout files, we can separate each error type into a separate cell, so a user can, for example, examine all metal-to-metal separation errors at once without worrying about other checks.

Furthermore, we would like to keep track of pairs of edges that violate each kind of check, so we can present the user with the specific edges that were in error for each error, although the specific edges involved are usually obvious to a designer.

Because we flattened the hierarchy before doing the DRC, we are able to check the entire cell with confidence. However, if the cell contains an array of 100x100 cells and there is one error in that cell, the user will get 10000 error messages reporting that error. Although this conforms to our desire to be conservative, possibly reporting more errors than there really are in the circuit, this massive number of extra errors is distracting and a user may miss a genuine error amid the mass of duplicate messages. Some scheme for reducing duplicate messages is a useful, though not vital, enhancement.

Ambiguous Corner Checks

Because we deal with edges and not complete polygons, we must determine for each pair of edges whether to compare them for potential width violation or potential spacing violation. The choice is obvious to a person viewing the layout, but to a program that is only looking at a pair of edges, it may be difficult to determine unambiguously.

In the case of the following figure, there is no ambiguity: the two edges overlap for some distance, so the check can be determined from the directions of the edges.

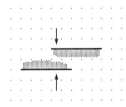

However, in the the following case, where the two edges do not overlap, we could be faced with either of the cases on the right. The spacing from corner to corner may be either the width of the polygon or the spacing between two parts of the same polygon or even between different polygons.

Layout Analysis

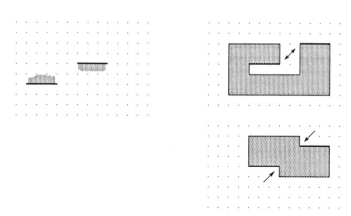

Of course, there is no ambiguity when the two edges are from two separate polygons -- it cannot be a width check because you can only have width checks between edges on the same polygon.

There is no way to determine absolutely from the two edges whether to do a width check or a spacing check. When we find this case, we can examine the other edges on the polygon to determine whether or not the area between the two corners is filled. We can do this using an algorithm similar to the scan line algorithm. We pick a point between the two corners and examine edges that cross the scan line at that point. If the point is in a filled area, we do a width check assuming the bottom case, if the point is in an open area, we do a spacing check. If the polygon is convoluted, as is the case in the following figure, we may do an improper check, but we will do the correct check on the intervening edge and the error, if any will be flagged. However, we may produce a false error message.

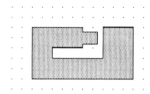

Another solution that has worked in this situation is to resize the layer by half the difference between the width and spacing distances. Then it doesn't matter if it is a width or spacing check, since they check the same distance.

Glitches

We have already discussed the major problems with edge-based analysis algorithms and their solutions. This section discusses briefly some of the less common, if not less important, problems.

Flagging Errors On Acute Angles

In our discussion of the selection of the edges to check, we said that we do not check rules between edges that are perpendicular. We catch violations when we check the edges connected to the perpendicular edge. The same applies for obtuse angles. We do not check spacing between edges that make an obtuse angle.

We must check rules between edges that make an acute angle. Clearly, case (a) is an error because the two edges are too close. We must also flag case (b) where the two edges touch. However, if the angle is close enough to a right angle the error probably should be ignored. This is the case when we have an 89 degree angle like case (c). The question that remains is how acute must the angle be before we flag it as an error? Generally, angles greater than 45 degrees are considered legal.

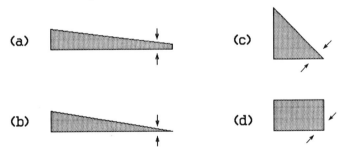

Coincident Edges

The examples in this chapter are rather simple in order to facilitate explanation of the algorithms. However, we do not always get such simple cases in real life. One common occurrence is the presence of many edges at the same point or even many edges superimposed on one another. The order we process those edges can change the output we generate.

For example, assume we are doing a merge operation and we encounter the case in the following figure where the two rectangles touch. As we progress up the scan line at location p, we set the wrap number to 1, then we encounter the two edges at the same location. If we see the edge from rectangle **A** first, then we set the wrap number back to 0 and write the portion of **a** that was a border. Then we see edge **b** and the wrap number changes back to 1, so set the new border. The result is that we don't merge the edges, we preserve the break between the edges. If we had found edge **b** first, then the merge would have worked correctly. On the other hand, if we are looking for the intersection area of the two rectangles, we would like to avoid finding the zero-width sliver of coincidence. The proper order for intersection is to find edge **a** before edge **b**.

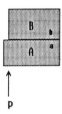

The conclusion is that we cannot process the edges in the same order for all edge operations. When we look through the edges in the scan line, we calculate everything that happens at that location and treat it as one event. After examining all edges at the coordinate, we update wrap numbers and write edges.

The problem is not one that applies only to parallel edges. We can have the same problem at a single vertex with many edges entering and leaving the point at once. When ANDing the two layers in the following figure, we would like to avoid reporting an overlap area at the single point of intersection on the left, though we must not miss the end of the overlap area on the right.

 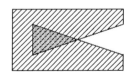

Technology File

We have covered the checking algorithms, but not how we describe the checks that we wish to make. As we have seen, some checks can require several steps, expanding layers and subtracting one from another. We simplify the description by defining a design rule language. The description of the rules for a technology are stored in a *technology file*. The file includes definitions of derived layers and specifications of checks. We mix the two kinds of statements to improve legibility. The definitions of transistor and contact layers are relatively straightforward. They contain and, or and not (&, | and -), and other operations, such as the crucify.

```
derive trans  dif & poly
derive pmc   poly & cut & metal
```

There are three kinds of checks: width checks on a single layer, spacing checks between layers and enclosure checks between two layers. The width check gives the minimum size and the name of the rule that was violated:

```
width dif 2  "diffusion minimum width"
```

For layer-to-layer checks, we name the layers, the separation distance and the name of the rule we are checking for error reporting:

```
spacing dif dif 3  "diffusion-diffusion spacing"
spacing trans cut 2 "transistor-cut separation"
```

And for enclosure rules, we list the inside layer, outside layer, the amount of overlap and the rule:

```
enclose cut metal 1 "metal enclose contact"
```

When we run the DRC, we read the technology file, create derived layers and build tables for the width and spacing checks. We do a second pass to check the rules.

Technology File for MOSIS Rules

Here are the rules for the MOSIS single-level-metal CMOS process we showed at the beginning of this chapter. The operations include none of the merging or re-sorting. We assume that those operations are done whenever needed.

```
# derive transistor and contact areas
derive trans   poly & dif
derive pmc     poly & metal & cut
derive dmc     dif & metal & cut

# simple width checks
width dif    2    "diffusion minimum width"
width poly   2    "poly minimum width"
width metal  3    "metal minimum width"
width cut    2    "cut minimum width"
width well   6    "well minimum width"
width trans  2    "transistor minimum width"

# check contacts exactly 2x2
derive nocut size cut -1
width nocut 99999   "transistor maximum width"

# simple spacing checks
spacing dif dif        3    "diffusion-diffusion spacing"
spacing poly poly      2    "poly-poly spacing"
spacing metal metal    3    "metal-metal spacing"
spacing cut cut        2    "contact-contact spacing"
spacing well well      6    "well-well spacing"
spacing trans trans    2    "transistor-transistor spacing"
spacing trans cut      2    "transistor-contact spacing"

# check poly diffusion space
derive t1      crucify trans 1
derive pcheck poly & -t1
spacing dif pcheck 1   "poly-diffusion spacing"

# simple enclosure checks
enclose pmc poly   1    "poly enclose contact"
enclose dmc dif    1    "diffusion enclose contact"
enclose cut metal  1    "metal enclose contact"

# diffusion well rules
derive dinw dif & well
enclose dinw well 6    "well enclose diffusion"
derive doutw dif & -w
spacing doutw well 6   "well-diffusion spacing"

# check poly overlap trans
derive t1         crucify trans .1
```

```
derive t2        t1 & poly
derive t3        crucify t2 2
derive t4        size t3 -.1
enclose t4 poly 0   "poly overlap transistor"

# check dif overlap trans
derive t1        crucify trans .1
derive t2        t1 & dif
derive t3        crucify t2 2
derive t4        size t3 -.1
enclose t4 dif 0   "diffusion overlap transistor"

# check contacts in select
derive cinsel    cut & select
enclose cinsel   select 1  "select enclose contact"
derive coutsel   cut & -select
spacing coutsel select 1   "select-contact spacing"

# check select overlap inside-dif
derive t1        dif & sel
derive t2        crucify t1 .1
derive t3        t2 & select
derive t4        crucify t3 2
derive t5        size t4 -.1
enclose t5 dif 0   "select enclose diffusion"

# check dif out of select to select
derive t6        crucify t1 2
derive dscheck   dif & -t6
spacing dscheck select 2  "select-diffusion spacing"

# check select overlap trans
derive t1        crucify trans .1
derive t2        t1 & select
derive t3        crucify t2 3
derive t4        size t3 -.1
enclose t4 dif 0   "select overlap transistor"

# check trans out of select to select
derive t5        crucify t1 3
derive tscheck   trans & -t5
spacing tscheck select 3  "transistor-select spacing"
```

Fast Sorting For Edge Files

All the operations we have discussed output edges when the maximum X value passes the horizon line. The output files are no longer sorted by minimum X, but they are not completely unsorted. They are sorted by increasing maximum X. If we must process one of these files, we can read the file backward and process right to left with minimal change in the algorithm.

We need only sort an edge file if we must combine two files that are sorted from different directions. Because the file is already sorted by maximum X, we can re-sort by minimum X in one pass using an algorithm developed by Szymanski and Van Wyk (1983). We read the edge file backwards getting edges in decreasing maximum X, and build a priority queue. We write out any edge that has a minimum X value greater than the maximum of the current edge we are reading from the file. The result is a file sorted by decreasing minimum X -- exactly the opposite order from the order we want. We read the file backward to process the edges in increasing minimum X.

These sorting improvements yield performance enhancements if the operating system's penalty for reading a file backward is small. However, if that penalty is greater than ten or twenty percent, the reading penalty overwhelms the savings from sorting, and we are better off sorting the data so we can read it forward.

Circuit Extraction

The major operations of circuit extraction are finding transistors, finding nets between the transistors, and accumulating areas and perimeters for making capacitance estimates. The result we want to write is a list of transistors with the nets on their source, drain and gate enumerated and a list of capacitors from the nets to ground or between nets.

Finding Transistors

In MOS circuits, transistors are created by the intersection of polysilicon and diffusion. The intersection code we have examined in the DRC will identify all transistors. We can uniquely identify them using the polygon numbering algorithm we saw earlier. Each polygon in the transistor layer is a separate transistor. Different types of transistors are indicated by the presence or absence of implants and wells, so we use the implant and well layers to determine which kind of transistor is represented by each result polygon on the transistor layer.

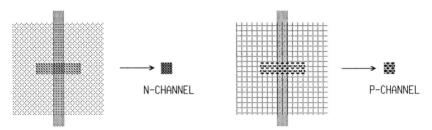

Finding Nodes

We want to label each edge with the conducting layer of the electrical node to which it belongs. The algorithm is a fairly obvious generalization of the polygon numbering algorithm we discussed earlier. Node identification associates a unique node number with each node in the circuit.

Connectivity is preserved over a continuous run on a wiring layer and by a contact connecting two layers. We process all edges on wiring layers simultaneously with the contact layers. When we encounter a contact, we *merge* the nodes on the layers that have filled area intersecting contact area. We cannot use the diffusion layer for connectivity, since a transistor breaks connectivity. We create a new layer for node finding, the diffusion minus the polysilicon layer (or, equivalently, but more efficiently, diffusion minus transistor). The result of the node identification is an edge file for the wiring layers with correct node numbers on all edges.

Circuit Extraction

We now make one pass simultaneously through the polysilicon, transistor and diffusion-transistor layers. As we read edges from the transistor file, we make transistor records identified by their transistor polygon number.

Layout Analysis

We keep polysilicon and diffusion edges in an active list. The horizon line is the lowest X value of any transistor. When we find a polysilicon edge that includes an edge of a transistor, we set the gate of the transistor to the node of the polysilicon edge. We treat the source and drain of a MOS transistor identically. When we find a diffusion edge that includes an edge of a transistor, we set the source of the transistor to the node of that edge. If the source is already set, we set the drain to that node. If both are already set, we have a multiple-source-drain transistor, such as those below. For simplicity, we add a node that represents the area under the gate and convert such structures into multiple two-terminal transistors.

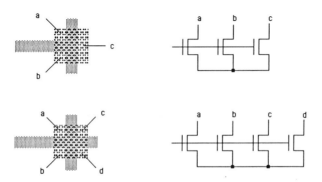

When the scan line passes the highest edge of a transistor, we write out a transistor record with the node names for its gate, source and drain. When the horizon passes a polysilicon or diffusion edge, we remove the edge from our active list. There is no need to write it out.

The operation of matching node edges and transistor edges requires that we consider vertical edges as well. As with the DRC, we generate vertical edges on the fly as we read the file.

Extracting Capacitance

We encounter three kinds of capacitance during extraction: area capacitance to the substrate, perimeter capacitance to the substrate and capacitance between parallel wires. Area and perimeter capacitances are much larger than capacitances between parallel wires, so we concentrate on them.

Capacitance is measured in picofarads per square micron for areas and picofarads per micron for perimeters. Each wiring layer and each transistor type has a scale factor that depends on the fabrication technology. We keep the scale factors for all capacitances and resistances in a technology definition file. We accumulate the areas and perimeters for the different types of material separately, multiply by the scale factors and total the result for the capacitance of a node.

Capacitance accumulation can be done as part of the polygon merge or as part of a separate pass over the edges. We maintain in the active list an association between edges that form the upper and lower border edges of the polygon. These edges are called *paired edges*. When one of the edges is to be written, or when the edges are no longer paired because of intervening edges, we determine the area between the two edges and total that area in the node record. We then update their left ends to the current scan line location for future area calculations. Since all line segments are straight, the areas are trapezoidal areas.

Perimeter distances are the lengths of the edges that are written from the polygon merge. The only difficulty in the perimeter calculations is that we must include border vertical edges in the calculations.

Transistor Aspect Ratio

The drive power of a transistor is determined by its *width* and *length*. The width is the length of the diffusion edge, the length is the length of polysilicon edge. The ratio of width to length is called the Z-ratio. It is inversely proportional to the resistance of the transistor when it is conducting. Many simulators use the Z-ratio to determine the performance of the circuit when it is driving its load. Therefore, if your simulator uses that information, you should extract it also.

We estimate the width of the transistor as half the diffusion border of the transistor and the length as half the polysilicon border. This estimate gives the correct value for rectangular transistors, but is less accurate for exotic transistor shapes such as bent or annular transistors.

Extracting Resistance

Wiring also has resistance. An accurate simulator will use the resistance of a wire to determine the shape and the speed of a signal propagating down that wire. Wire resistance is not always an important design consideration at this time, but analysis shows that it will become more important in the future.

The resistance of a transistor is the inverse of the Z-ratio we discussed, with a technology-dependent scale factor. However the resistance of a wire is not at all as simple as the capacitance calculation we described earlier. The major difficulty is that we must determine the direction of current flow. Of course, current flows between transistors, but the paths may be complicated.

To estimate the resistance of wires, we determine which edges form wires and calculate the length to width ratio of those wires. The first step is to find long parallel edges on the wiring layers that enclose layout. If parallel edges are far apart or if they are very short, then we are not interested, because the resistance will be very low.

We make the assumption that long parallel lines on wiring layers are resistive wires, so we output a new layer, the resistor layer, for the parallel edges we find. To separate nodes, we subtract the resistor layer from the wiring layer. We can now perform the node identification step, and when writing transistors, we write resistors as well. The resistance of the resistor is the length of the parallel lines divided by the distance between them times the resistance scale factor for the layer.

This algorithm does not treat corners accurately, nor does it handle non-rectangular shapes. There are algorithms for dealing with the shapes very accurately by doing a detailed analysis of the current flow on a case-by-case basis. They can be very involved and we will not discuss them here. Readers are directed to the work of Bastian, et al (1983), Horowitz and Dutton (1983) and McCormick (1984) for more information about resistance extraction.

The resistance algorithm we described generates two resistors and three nodes for a jogged wire. It is advantageous to us to reduce the number of nodes and resistors by combining serial resistors, totalling their resistance and omitting intervening nodes.

We have also neglected the case of RC lines. Wires on integrated circuits have resistance and capacitance distributed along their length. We would like to output the resistance and capacitance in lumps, but to do so incurs problems with accuracy. A reasonable option is to output half the capacitance of the wire on each side of the resistor, as shown in the following figure. If the resistor is very large, it is more reasonable to divide the wire into more segments. An option is to split transistors that are greater than some preset size.

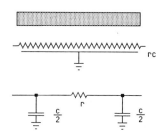

The Netlist Data File

As we stated previously, the output of the circuit extractor is a list of transistors with the transistor type, the gate, source and drain nodes. In addition, we include statements for capacitors, resistors and nodes. The following BNF describes a netlist file:

file	= fileHeader cellHeader body endStmt
fileHeader	= "V" versionName ";"
cellHeader	= "CELL" cellName technologyName netlist ";"
body	= { (transistorStmt | nodeStmt | capacitorStmt | resistorStmt | aliasStmt) ";" }
transistorStmt	= "T" ("n" | "p" | "d") gateNode sourceNode drainNode x y w l
nodeStmt	= "N" capacitance name
capacitorStmt	= "C" capacitance node1 node2
resistorStmt	= "R" resistance node1 node2
aliasStmt	= "A" { node }
endStmt	= "E"

In the netlist file, the nodes are identified by a name which may be the node number we determined during extraction. The transistor has a type: n for n-channel, p for p-channel and d for NMOS depletion. The transistor also includes the names of nodes to which it connects, its location to assist the user to identify the transistor during simulation, its width and length. The alias statement lets us name nodes.

Options

Optional Capacitance and Resistance

Not all simulators use all the information our extractor can generate. Extraction of capacitance and resistance is expensive and enlarges the data file considerably. We make extraction of capacitance and resistance optional so users who want a simpler simulation do not pay the penalty of a complex extraction.

Node Names

In order for users to make sense of their simulations, they will have to refer to signals by names, rather than by our arbitrarily-generated numbers. The user-defined names should override the machine-generated names in the data file.

Node names come from the connectors in the layout or from explicitly-placed node names in the layout, a feature we would have to add in the layout editor. Either way, we must modify our node determination step to check for predefined names and include them in the output netlist file as alias names. Hierarchical designs may have many nodes with the same name. To avoid confusion, we ensure that all node names are unique in the final netlist by building hierarchical names. A hierarchical name has a prefix that consists of the names of all the instances down the hierarchy in which that node resides. The name of the node OUT in an instance latch1 which is itself an instance of clkctr will have the name clkctr.latch1.OUT.

Omit Checking in Areas

In both DRC and circuit extraction, we encounter areas that should not be processed. These areas are usually designers' initials or company logos, but may be special circuits or experimental structures that violate the usual rules or cannot be simulated correctly. We indicate that these layers are not to be checked by providing two additional layers in the layout editor, the NOCHECK layer that indicates the area is not to be design rule checked, and the DISCONNECT layer that indicates the area is not connected to the circuit. The user covers the special structures with these layers to avoid false errors.

In the initial stages of the DRC and extractor, we subtract the NOCHECK or DISCONNECT layer from all other layers in the cell, so those areas are simply not part of the calculation. However, when we remove areas in this manner, we may introduce design rule violations by NOCHECKing part of a rectangle leaving a rule violation on the remainder.

Hierarchical DRC and Extract

The DRC and circuit extraction we described are flat. The hierarchy is removed, so we deal only with edges, no instances. Although flattening makes the algorithms simpler, we create a huge amount of data to process. In addition, if we use a cell a hundred times and we find a DRC error, we report one hundred errors, when in fact the user only made one error. These false errors make the error reports difficult to read, so some true errors may go unnoticed.

Whitney (1981) and Scheffer and Soetarman (1986) proposed hierarchical solutions to these problems. We can first check the cells, then check their assembly without re-checking the interiors of the cells we have examined. We perform these checks recursively until we have checked the largest cell. Since we don't check all the layout of all the subcells at once, we have much less data. In addition, we check each cell body only once, so we report each error only once.

However, hierarchical analysis can lead to additional work. We check each cell, and we check that cell as it borders other cells. So we check the edges of a cell once for its definition and once for each environment in which it is used. In addition, if the user puts some additional layout across the top of a cell, we must re-check the entire cell with the new layout.

If we are extracting a circuit, we must be careful that nodes that include layout touching the cell are connected to the cell. We re-extract the cell for every instance where layout crosses the cell because that new layout may add or remove transistors and may change capacitances or resistances on nodes inside the cell. And, of course, we will have to modify our netlist file format to include instances and connections to nodes in them.

In general, hierarchical DRC and extraction are much more difficult than they might seem on the surface. Therefore, most analysis software works on flat layouts addressing performance problems with efficient algorithms and addressing multiple-error problems with post-check filters. There are hierarchical analysis packages, but they usually require some restrictions on the layout to be effective. Barring layout on top of an instance is a severe restriction because it is a preferred technique for layout generators. However, if the performance of layout analysis is sufficiently improved by such a limitation, tool makers and designers alike may accept the restriction.

Mask Tooling

The job of mask tooling is to convert the layout the user has drawn into the masks that the fabrication facility needs to manufacture the chips. This job is more than a mere format conversion. Typically, the layers that users draw are not one-for-one equivalent to the layers that the manufacturing facility needs. Some layers are the AND of two or more layers, some layers are the inverse of what has been drawn, most layers will have to be bloated or shrunk to offset manufacturing realities.

The operations we have described with design rule checking are the operations we need to perform all mask tooling operations. Therefore, the software we have developed to handle edge files is capable of performing the mask tooling as well. We write a new design rule check file to generate the tooling. In mask tooling we are more concerned with accuracy than we were in checking. The spikes from bloating and shrinking polygons that were annoying in the DRC are fatal in mask tooling.

Typically, the output for mask tooling is a list of trapezoids that describe the layout. Conversion of edge files to trapezoids is left as an exercise for the reader.

Other Algorithms

The edge-based algorithms presented in this chapter are the most popular, general and successful algorithms currently in use. However, they are not the only techniques for layout analysis. This section briefly outlines two other techniques: trapezoid-based and raster-based.

Trapezoid-Based Algorithms

We had some problems when checking widths and spacings of edges with cases where we didn't know whether to check a width rule or a spacing rule. We had that problem because we didn't know which side of the shape was really an inside area. We can resolve this problem by keeping the layout as trapezoidal areas rather than as edges.

We keep paired edges in many of the edge based algorithms. A trapezoid is a pair of edges with implied vertical edges. We can clip trapezoids to avoid re-sorting and we can simultaneously resize and merge. A trapezoid has a definite inside and outside, so we no longer have that ambiguity. In addition, trapezoid files tend to be smaller than edge files. Trapezoid handling is efficient, because most of the cases are rectangles.

There are disadvantages of trapezoids, also. Since they do represent subsets of the area, we cannot simply check the width of a trapezoid to find a width violation, we may have to examine nearby trapezoids. The task of searching nearby features will slow down the algorithms.

Raster-Based Algorithms

A significantly different method of checking layout is to convert the chip to be checked to a raster just as if we were going to output to a raster plotter with a coarse resolution (Baker and Terman 1980). A typical resolution is about 1 lambda, but depends on the size of the rules to be checked.

We examine very small areas for violations. For a 2λ rule, we check a 3x3 square of pixels. Circuit extraction is even easier, since we need to check only three pixels per layer at one time to determine connectivity.

Raster conversion is very simple and fast. It consumes a lot of memory, but we only need convert small pieces at a time. However, raster DRC does not work well if layout does not fall on a coarse grid or if there are non-orthogonal lines. To handle these cases, we must use a finer grid. A finer grid means more memory and the checking window must be correspondingly larger. It was easy to list all the cases where a 2-lambda rule would be violated in a 3x3 window, but it would be impractical to list the ways a 4-grid rule would be violated in a 6x6 window, or even a 5x5 window. There are simply too many cases for this to be practical.

For similar reasons, larger design rules, those requiring spacing of 4 or 5 lambda are extremely time consuming to check. These problems have reduced the practicality of raster-based layout analysis.

Exercises

Programming Problems

1. Write a translator from layout files to edge files.

2. Write a polygon merge program that reads and writes layout files.

3. Write a package that performs boolean operations on mask layers. Use your package to implement a design rule checker for the rules given in this chapter.

4. Write a translator from edges to trapezoids.

Questions

1. Estimate the disk space required to store a design as objects and as edges. How much space would be required if you could exploit hierarchy?

2. What sorting algorithm would you choose to sort edges initially? Why?

3. How would you incorporate sorting into the edge operations so we can do it in parallel with other operations? How many edges would you expect to keep in memory at once? How many in the worst case?

4. What are the advantages and disadvantages of keeping edge files as binary versus text?

5. We have described the edge-based sorting algorithm working with a vertical scan line. Are there any advantages or disadvantages to using a horizontal scan line instead?

6. Because we flattened the hierarchy before doing the DRC, we are able to check the entire cell with confidence. However, if the cell contains an array of 100x100 cells and there is one error in that cell, the user will get 10,000 error messages for the one error. How would

you address this problem? Suggest at least one programming solution and one methodology change.

7. How would you modify the circuit extractor to include capacitors between nodes where the wires cross one another?

8. Describe an edge-based algorithm to calculate capacitance between parallel wires. Hint: use an algorithm similar to the resistance extraction.

9. We extract capacitance on a node by totalling the areas and perimeters for different layers, multiplying by scale factors and totalling the result. What are the advantages and disadvantages of skipping the scaling and writing the individual areas and perimeters for each node to the netlist data file?

10. Describe an edge processing algorithm that checks that every N-well has at least one well contact. A well contact in the N-well is a contact inside select in the well. List the derived layers you need.

11. How does symbolic layout simplify DRC and extract?

12. What problems do you foresee if you DRC a file that does not circuit extract properly? What problems do you foresee if you extract a file that does not DRC correctly?

References

Edge-Based Algorithms

J.L. Bentley and T.A. Ottmann, "Algorithms for Reporting and Counting Geometric Intersections", *IEEE Transactions on Computers*, vol. C-28, No. 9, September 1979.

H.S. Baird, "Fast Algorithms for LSI Artwork Analysis", *Proceedings of the 14th Design Automation Conference*, 1977.

U. Lauther, "An O(N log N) Algorithm for Boolean Mask Operations", *Proceedings of the 18th Design Automation Conference*, 1981.

L.K. Scheffer and R. Soetarman, "Hierarchical Analysis of IC Artwork with User-Defined Rules", *IEEE Design and Test*, February 1986.

T.G. Szymanski and C.J. Van Wyk, "Space Efficient Algorithms for VLSI Artwork Analysis", *Proceedings of the 20th Design Automation Conference*, 1983.

T. Whitney, "A Hierarchical Design Rule Checking Algorithm", *Lambda*, First Quarter, 1981.

Symbolic and Raster Algorithms

C.M. Baker and C. Terman, "Tools for Verifying Integrated Circuit Design", *Lambda*, Fourth Quarter, 1980.

W.S. Scott and J.K. Ousterhout, "Magic's Circuit Extractor", *IEEE Design and Test*, February, 1986.

Resistance Extraction

J.D. Bastian, M. Ellement, P.J. Fowler, C.E. Huang and L.P. McNamee, "Symbolic Parasitic Extractor for Circuit Simulation (SPECS), *Proceedings of the 20th Design Automation Conference*, 1983.

M. Horowitz and R.W. Dutton, "Resistance Extraction from Mask Layout Data", *IEEE Transactions on Computer-Aided Design*, vol. CAD-2, No. 3, July 1983.

S.P. McCormick, "EXCL: A Circuit Extractor for IC Designs", *Proceedings of the 21st Design Automation Conference*, 1984.

CHAPTER 9

SIMULATION

An integrated circuit designer uses simulation for two tasks, to verify that the design he made is actually what he wanted to make and to determine the operating speed of the resulting circuit. The same simulator may be used two different ways: interactively, for initial debugging of a circuit; and batch-like, to exhaustively verify the circuit. Some of the features required for the different types of operation are different, so a simulator may include more than one way to set and observe the simulator state, among other things.

When we write a simulator, we trade off simulation speed against accuracy. The goal of the simulation writer is to choose a representation of the problem that will give accurate enough information while running fast enough. The definition of "enough" varies among designs, among different stages in a single design and among designers. Therefore, several different simulators may be necessary in a single design system. In this chapter, we investigate simulators for behavior, logic and transistor switch-level timing.

As with the DRC, when we trade away accuracy, we take care that we err on the side of conservatism. We want the simulator to be pessimistic about the performance of the circuit, when that is being measured. Users are more tolerant of a simulator that gives results poorer than their actual circuit than they are of one that gives overly optimistic results.

Knowledge of the domain to be simulated is essential for any simulator writer. Although the concepts specific to simulation are presented here, this chapter assumes some knowledge electronics and integrated circuit design.

Types of Simulators

Types of Descriptions

There are many different kinds of simulators, differing primarily in the precision with which they represent the actual device. The most precise simulators are *circuit simulators* which model transistors accurately and calculate voltage and current on all electrical nodes through small time quanta. A circuit simulator numerically solves the differential equations for all nodes simultaneously. The result is an accurate, though slow, simulation.

We can also simulate a circuit as transistors connected with electrical nodes, modelling transistors as switches and nodes as capacitances. This type of simulator is a *switch level* simulator. If we include delays due to capacitance and transistor drive resistance, it is called a transistor *timing simulator*.

A *gate-level simulator* or *logic simulator* simulates logic gates such as latches, AND gates and OR gates. The simulation does not model the physical transistors accurately, but many circuits, particularly those generated automatically, are composed of gates built of transistors. Since we model the actual form the designer specified, we can provide better feedback. Further, the simulation accuracy may be improved because we can assign gate delays that we measure from the silicon, avoiding the approximations due to a transistor model.

A *register transfer level* simulator, or *RTL simulator*, operates on numerical values with arithmetic operations representing functions. It is usually used to verify a processor at the functional level, describing operations in terms of the registers and ALU operations, hence the name.

A *behavioral simulator*, simulates behaviors of blocks of logic as pieces of code. Basically, a user writes small procedures that model the behavior of elements such as ALUs or whole processors. These procedures set the values of global variables to represent nodes in the circuit.

Signals

We use the term *node* to refer to a wire or connection in the simulation. Each node has a signal on it. A *signal* is the unit of information in the simulation. We simulate by setting signals on nodes.

The first step in any simulation is to determine what values we give our signals and, correspondingly, what the operations are on those signals. Complex simulators model signals as currents and voltages. Although this representation is very accurate, it also leads to a very slow simulator. At the other extreme, we can model signals with truth values of *true* and *false* or 0 and 1. These values may correspond to 0V and 5V, or there may be some other voltage range. We simplify the problem by dealing with only discrete states of the signal. This simplification eases the calculation we must do to determine the resulting signals, and presents more precisely the information a digital designer may need. However, it makes the simulation less accurate, so the information may be less reliable.

In a behavioral simulator with primitive operations of addition and multiplication, a single signal may represent a multiple-bit bus. A user may set the value of all bits on the bus by a single integer assignment. It may also be convenient in a behavioral simulation for a signal to be a string indicating the op-code of a function in a processor, for example.

A signal may include other information beyond its functional value. We may have states that have no physical realization. For example, the simulator may not know what the value is on a node if the node is uninitialized or driven to two values simultaneously. In that case, we have a special value to represent *unknown* or *error* signals.

Time

The accuracy with which simulators model time varies with the detail of the simulation. A behavioral simulator may simulate without considering the timing of the signals it passes. On the other hand, we may take great pains to ensure that all signals are scheduled to arrive at their destinations at times that are consistent with the design we made. Accurate timing slows down the simulation.

A Simple Behavioral Simulator

One obvious way to simulate a function is to write a program to calculate the function. We write procedures which we call *functions* or *functional models* to calculate the operations of the cells in the chip. We represent signals with variables in our programming language. Each procedure takes some variables as inputs and modifies some variables as outputs. Therefore, we have all the power of the programming language to calculate the functions and we have any data type or structured data we want to represent the signals.

For example, suppose we have a processor that looks like this:

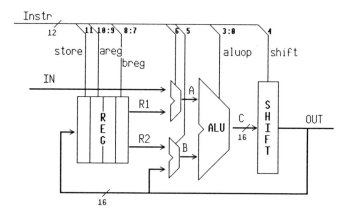

We can write the simulation for this processor like this:

```
procedure alu(integer a,b;  string aluOP;
  produces integer out);
begin
  if aluop="add" then out := a+b
  else if aluop="sub" then out := a-b
  else if aluop="passA" then out := a
  ...
end;

procedure processor(integer in,instr; produces integer out)
begin
  string aluop;
  boolean shift,store;
  integer areg,breg,aluout;
  integer array (1 to 4) registers;
  own integer abus,bbus;

  # decode the instruction
  shift := extractField(instr,4,4)=1;
  if extractField(instr,3,0) = 0 then aluop := "add"
  else if extractField(instr,3,0) = 1 then aluop := "sub"
  else if extractField(instr,3,0) = 2 then aluop := "passA"
  ...
  areg := extractField(instr,8,7);
  breg := extractField(instr,10,9);
  store := extractField(instr,11,11) = 1;
```

```
# run the processor
if extractField(instr,6,6)=0 then abus := in
    else abus := registers[areg];
if extractField(instr,5,5)=0 then bbus := out
    else bbus := registers[breg];
alu(abus,bbus,aluOP,aluout);
out := if shift then aluout*2 else aluout;
if store then registers[breg] := out;
end;
```

Time and the Event Queue

A simulation like the preceeding one is helpful for designing the overall system, but it becomes unwieldy when there are large numbers of things to simulate. In addition, there are some problems with timing and parallelism. These problems limit the accuracy of the simulation.

Let us take a close look at a very simple case to expose one of the major problems with this program-style behavioral simulation. Suppose we have a functional model for a NOR gate and we want to simulate a cross-coupled latch:

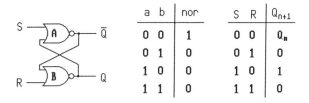

a b	nor		S R	Q_{n+1}
0 0	1		0 0	Q_n
0 1	0		0 1	0
1 0	0		1 0	1
1 1	0		1 1	0

```
procedure latch(boolean s,r;  modifies boolean q,qbar);
begin
  qbar := nor(s,q);   # gate A
  q    := nor(r,qbar); # gate B
end;
```

If we write the function for the latch like the code above, it doesn't always work. Suppose S, R and Q are 0 and qbar is 1. When the input S changes from 0 to 1, we must evaluate gate A before gate B so the code functions properly. But when Q is 1; S, R, and qbar are 0; and input R changes from 0 to 1, we must evaluate gate B before A. The code produces an incorrect result, setting both q and qbar to 0. One solution is to re-write the code so each gate is evaluated twice, so it doesn't matter which input changed. But it is possible to build arbitrarily-complex cases.

We are noticing the difference between a sequential program and a parallel integrated circuit. We need a more powerful mechanism to determine which signals to evaluate at what times. To separate the program control flow from the simulation control, we make data types for nodes and functions and make a linking mechanism for them that is independent of the programming language we use to evaluate the functions. We build a data structure that consists of functions and nodes. A function has a list of its input nodes, its output nodes and the code of the function. A node has a value.

```
class function(
  pointer(list) inputs,outputs;
  procedure eval(inputs,outputs);
);

class nodeClass(signal value, nextValue);
```

Now when we simulate, we supply values on the nodes of the inputs. We scan through a list of all functions, executing them and and updating their output nodes with new values. Since we want all functions to be evaluated with the same set of initial node values, we put the new value of each node into **nextValue**, a temporary place in the node description. When we have finished evaluating all functions, we copy those values onto the nodes in the circuit. Then we are ready to begin the next cycle and evaluate functions again. As an obvious improvement, we can mark which nodes changed and only evaluate functions that take input from changed nodes.

In this kind of simulation, all evaluations conceptually take place at once. A signal change propagates through one function each time we make a pass through the functions. Therefore, we can count the number of passes we make over the functions and use that as a measure of time. Since a signal passes through one function each time step, each function has one unit of delay. This kind of simulation is known as a *unit delay simulation*.

We know, however, that not all functions take the same amount of time to perform. Therefore, we would like to set nodes to their new values at some time in the future and evaluate each function only when an input changes. A simulator that works like this is called an *event-driven simulator*.

In an event-driven simulator, when we determine that we must change the value of a node, we schedule an *event*. An event includes the node to change, its new value and the time at which that change takes place. We keep all pending events sorted in time in a priority queue called the *event queue*.

```
class eventClass(
  pointer(nodeClass) node;
  signal value;
  integer time;
);
```

We take the next event from the event queue and set the current time to that time. We set the value of the node in the event and evaluate every function that takes the node as an input. The function evaluations generate more events which we add to the event queue.

With events generated in this manner, each cycle of the simulator only invokes those functions that actually have to be re-evaluated. Since very few of the functions are invoked on each event, we can avoid scanning the list of functions by keeping with each node a list of the functions it drives. However, if several inputs to a function change simultaneously, we wind up evaluating the function several times when only one is needed. To avoid this problem, we take from the event queue all simultaneous events, mark which functions are to be re-evaluated for those changes and evaluate all functions in one pass. Each function is evaluated at most once per event time and the results won't be dependent on the order of simultaneous events in the event queue.

```
currentTime := nextEventTime;
while nextEventTime=currentTime do begin
  e := nextEvent;
  for all functions (f) with e.node as input do
    putInList(evalList,f);
end;
for all functions (f) in evalList do evaluate(f);
```

Modelling Shared Busses

We have made an assumption so far that every function always drives all its outputs and that every node has only one function that drives it. This may not be the case. Large structures frequently contain shared bus structures: any one of a number of functions can write to the bus; others may disconnect themselves from the bus. We must check conflicting writers on the bus and re-evaluate the bus when one writer stops driving it. The bus itself acts like a separate function. We can make the function explicit by requiring users to wire all bus connections to a separate bus function or we can find such structures when we read the input and generate the bus functions ourselves.

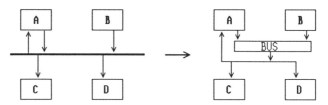

There is one more piece needed to evaluate the bus, that is some way to represent a signal that is not driven, the high-impedance state. When a function is not driving the bus, its output goes to a high-impedance state, not driving the bus. To model these busses, we have a signal value *high-impedance* in addition to *true* and *false*.

A Logic Simulator

A *logic simulator*, also known as a *gate-level simulator* (d'Abreu 1985) simulates a circuit composed of logic gates: AND, OR, NAND, NOR and INVERT. A logic simulator is simpler than the behavioral simulator we just saw because there is only one kind of signal, a logic level. The functions are all from the preceeding set of functions and can be built-in and optimized.

Logic Signals

In our simulation, we will use integer 0 and 1 to represent *false* and *true* logic levels. We add a third signal value, *unknown*, to make a *three-state simulator*. We use the integer 2 to represent *unknown*, but refer to it as X. We set a node to *unknown* if we cannot determine its actual value. When we start the simulation, we don't know the value of internal nodes. Rather than choose to set them all to 0 or 1, all nodes are set *unknown*. As the simulation runs, known values propagate into the circuit. If some nodes never become known, we have identified a situation where the initialization is inadequate. Users can supply *unknown* inputs to test that the circuit functions without regard to some inputs.

Simulation

The third state doesn't cost us much in performance, since we can cast a boolean operation as a matrix and index into the matrix to find the result. The third state merely enlarges the matrix. The following is the matrix for AND. ANDing a 0 with a X results in a 0, since it doesn't matter what the unknown value really is, the result will be 0. We have similar functions for other gates. We can add a new function by defining the outputs it generates for each set of inputs. Other functions common in logic simulators are decoders, multiplexers, latches and flip-flops.

	0	1	X
0	0	0	0
1	0	1	X
X	0	X	X

Data Structures

The **node** record in the simulator has a current value, a list of cells to which it connects as inputs, and a capacitance, which we use for timing.

An **instance** record contains a cell type, which determines its function, the delay for the cell and a list of nodes on its outputs. Although all the logic cells we mentioned have only one output, most logic simulators allow latches as primitive cells.

As discussed previously, an **event** includes a time, a pointer to a node and a new value for that node.

Evaluating a Cell

When we evaluate a cell, we set the new value of its output node or nodes at some time in the future. The procedure **evaluateCell** takes as input the cell to evaluate and the current time. Since each cell takes a different amount of time to generate its output, the piece of code that sets the output value also sets the time of the event for that output to the current time plus the delay of the gate.

```
procedure evaluateCell(
  pointer(cellClass) c;
  integer currentTime);
begin
  case c.cellType of begin
    [andGateCell] begin
      newValue := andTable[c.in[1].value,c.in[2].value];
      newTime := currentTime+andCellDelay;
      if c.output.value<>newValue then
        scheduleNewEvent(c.output,newValue,newTime);
    end;
    [orGateCell] begin
      ...
    end;
    ...
  end;
end;
```

In reality, the delay of a signal depends not only on the gate driving it but also on the capacitance it drives: the fanout of the gate and the capacitance on the wire. We would like the signal to be delayed not only by the normal propagation delay of the gate but also by the number of gates of fanout and the amount of capacitance on the wire. We can make this addition by setting newTime with a more complex function. For example:

```
newTime := currentTime + andCellDelay +
  andCellFanout*c.output.cellListLength +
  andCellCapacitanceFactor*c.output.capacitance;
```

We can now start playing games with accuracy, making signal delays that depend on which input changed or whether the signal is going high or low. We can model capacitance delays more accurately with higher-order terms in the equation. We are trading off accuracy of the simulation for the amount of time the simulation takes to run. It doesn't pay to be any more detailed than this with the time calculation, since the logic gates are an approximation anyway.

With the evaluation model we have developed, the output of a gate will change after the desired delay for each change in the input, no matter how frequently the input changes. In real circuits, an input signal must remain stable for some length of time for the gate to change. That length of time is called an *inertial delay*. If the duration of a signal is less than the inertial delay, the gate doesn't change. For example, suppose we have an

Simulation

inverter shown in the following figure, and we input the spike signal on its input. If the spike is long enough, the inverter will pass the spike. If the spike is too short, within the inertial delay of the inverter, the inverter will not be able to respond and will not change.

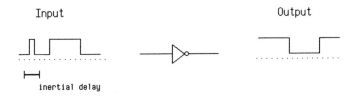

To add inertial delays to the simulator, we change the way we add an event to the queue. When we schedule an event for a node within the inertial delay time of another event we remove from the queue the pending event. If the inertial delay is equal to the delay for the new event, then every new event for a node removes all pending events.

Basic Simulator Function

As we discussed with regard to behavioral simulation, the event-driven simulator loop starts by removing an event from the event queue. We change the node indicated by the event and re-evaluate the gates that use that node. Re-evaluating those gates changes more nodes and we put the events for those changes into the event queue.

```
procedure simulate;
begin
  while eventQueue not empty do begin
    event := nextEvent(eventQueue);
    currentTime := event.time;
    event.node.value := event.value;
    for all cells, c, in event.node.cellList do begin
      evaluateCell(e.eventType, c, currentTime);
    end;
  end;
end;
```

Controlling and Observing the Simulation

So far, we have described the meat of the simulator, what we call the simulation *engine*. Around the simulation engine, we include interface facilities that allow a user to interact with the simulation, set node values, observe nodes in the simulation and modify the netlist. The commands to drive the simulation are very similar to the commands used to operate a powerful software debugger.

Types of Input Control

When controlling the simulation, users must supply *stimulus* values to drive the simulation and observe the value on any node. We want to be able to supply a stimulus to a node that performs in the simulation just like a node change that we encountered in the event queue. Setting a node by user command should work just like setting the node by the output of a function.

We also want to *force* a node to a value and keep it at that value. A node that is forced to a value cannot be changed by the operation of the simulator. We require this kind of force when we want to isolate part of the circuit to test it without interference from other parts of the circuit. Of course we must also provide a command to unforce the node.

A third kind of node setting is used to set a node to a periodic signal like a clock. Many circuits include clock signals, and an easy way to set up such a function is necessary to simulate them. A clock definition set the clock value (0, 1 or X) and how long it stays at each value. We assume that the signal begins at the current time and repeats when it reaches the end. For example, to set phi1 to 0 for 100ns then to 1 for 40ns, then back to 0:

```
clock phi1 0 (100) 1 (40)
```

The syntax may vary, and the clock definition may have more than two transitions. Like other nodes, clocks may be forced or normal.

Types of Output

A common form of output from a simulator is *trace output* to allow a user to *watch* a node. Whenever a watched node changes, we write out the time, the node name and the new value of the node. We also would like the option to output the value of a node periodically whether or not the node has changed.

Another form of output that is preferred by many users is waveform output. As its name implies, a *waveform* is a drawing of the signal value with time running horizontally and voltage (or 0, 1, X) vertically. It is a relatively simple translation from trace output to waveforms, and one that enhances the readability of the output considerably.

Traversing the Netlist

When a simulation uncovers a problem, a user needs facilities to follow problem nodes to uncover the cause of that problem. This feature is the equivalent of viewing the source code in a programming language. Since there is no source and since the circuit cannot be conveniently represented by a sequential list of commands, we have somewhat different debugging features.

There are four major commands to do view the netlist, **show instance**, **show node**, **show after** and **show before**. **show instance** takes the name of an instance and gives information on its type, delay and the nodes on its inputs and outputs. **show node** lists the node value and its capacitance. **show after** lists information on instances of which the node is an input. **show before** lists information on instances of which the node is an output. These commands allow a user to examine nearby nodes of a node that is in error. He can use subsequent commands to follow problems back to their source. In the following circuit, **show after** node D shows instances 2 and 3 with nodes E and F. **show before** shows instance 1 with nodes B and C.

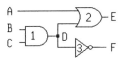

Altering the Netlist

When a user finds an error, he may wish to alter the netlist slightly to see if a proposed change would fix the error. The simulator interface includes the ability to add nodes, short nodes together and to add instances. These are the same operations that we do when we read a netlist, so the code of the user interface and the code of the netlist language parser may be the same. Users may also wish to split nodes, remove nodes, or remove instances from the simulation. Altering the netlist of a simulation that has been running is not without peril. Every node that is touched by a change must be re-evaluated; we must remove events from the event queue for nodes that do not exist anymore; and we must alter events for nodes that have been split.

Controlling the Simulation

To control the simulation, we need a facility to let the simulator run for a period of time, then return to the user interface. Our job is not so much invoking the simulation as it is stopping the simulation.

We stop the simulation by making a new kind of event, a *user interface event*. When we find a user interface event in the event queue, we execute the code that performs our user interface, asking for input from the user. We remain in the user interface while the user examines nodes and sets node values. When the user gives the command to continue, we schedule a new user interface event in the event queue and exit from the user interface.

To run the simulation for a specific amount of simulated time, we schedule a user interface event at that time in the future then exit the user interface. We can single step the simulation by setting a user interface event at the next clock time. We can also subdivide the clock period into *phases*, each transition of a clock begins a new phase. Then we can schedule a user interface event after a phase.

We can stop the simulation when a certain condition holds by combining the ideas of watched nodes with a user interface event. Watched nodes type their values when they change. We can watch a node and, instead of printing the value, schedule a user interface event as the next event in the event queue.

A conditional break point may depend on more than just the value of one node, it may be a combination of values. We can implement this complex conditional break by parsing the termination condition equation and evaluating it with an interpreter. A more elegant solution is to build more logic in the simulator to perform the complex conditional equation, then set a break on the result. We add logic that the user did not specify and which he should never see. The evaluation of the break point logic is done with the simulation engine, which slows the simulation somewhat, but the performance penalty is proportionally small when the circuit is large. However, this technique has no performance penalty with a *hardware accelerator*, a special-purpose parallel computer to perform the simulation. A hardware accelerator dedicates hardware to the instances in the circuit so all evaluations take place in parallel. The additional logic for the break points runs in parallel with the logic of the design.

Managing the Event Queue

The most common operations in the simulation are adding and removing events from the event queue, so techniques that optimize the event queue dramatically improve the performance of the simulator.

The event queue is a priority queue from which we remove the lowest element. A linked list with an insertion sort requires $O(n^2)$ time for updating the list. Traditional heap sort or other sorting methods require $O(n \log n)$ time. We minimize sorting by implementing event queue with a bucket sort that sorts in $O(n)$ time. We use a knowledge of expected delays to choose the number of buckets.

Time in the simulator is not continuous, but incremental. We simulate with a smallest unit of resolution, which we call a time quantum. Let us assume that a time quantum is 0.1ns. We define an array of event pointers called a *time wheel* (Ulrich and Suetsugu 1986). If the time wheel array is 500 entries long, we have one slot for each time quantum up to 50ns in the future. We choose 50ns because we expect that nearly all events will be scheduled less than 50ns in the future. Each slot in the wheel has a list of events, all of which happen at the same time.

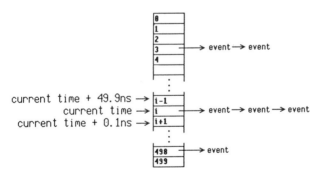

For events farther out in the future than 50ns, we keep an unsorted list of events called the *far list*. When we reach the end of the time wheel, we make one pass through the far list, removing all events that occur within the next 50ns and putting them in the proper places in the time wheel.

As we simulate, we advance the current time and always take the next event out of the current time slot. When we schedule a new event less than 50ns in the future, we put it directly in the time wheel, at location `(currentTime+newDelay) mod timeWheelLength`. Thus, most events are sorted immediately, while events that take longer and fall into the far list are sorted into the time wheel efficiently.

Simulating Transistors

Real integrated circuits are composed of transistors connected with wires. The circuit extractor in Chapter 8 produces a list of transistors and nodes. A MOS transistor does not behave like a gate, it behaves more like a switch, and it is important to model the bidirectionality of transistors in the simulation (Chawla et al. 1975, Bryant 1980 and Bryant 1984).

In this section we describe a *timing simulator* for a netlist of transistors. This timing simulator is based on the work by Terman (1983) and enhanced by Schaefer (1985). It models transistors as resistors and wires as capacitors. It takes into account the bidirectionality of transistors, a feature that our logic simulator did not. The timing simulator also takes into account implicit storage nodes.

It is possible to build a transistor-level simulator that does not take timing into account, and such a simulator will run faster than the timing simulator we investigate. However, circuit timing is essential for evaluation of integrated circuit correctness, and the timing features of this simulator can be removed to improve performance.

MOS Transistor Circuits

A transistor in a MOS circuit can be modelled as a switch. When closed, it connects the path from its source to drain. When a node is connected though a series of closed transistors to a power supply node, VDD (always 1) or VSS (always 0), the node takes on the value of the power supply. We use the terms VDD and VSS interchangeably with 1 and 0.

We treat three kinds of transistors, *n-channel transistors* that are closed when the signal on the gate is 1, *p-channel transistors* that are closed when the signal on the gate is 0 and *depletion transistors* used in NMOS circuits that are always closed, but have a relatively high resistance. A transistor that is conducting is said to be *on*, a transistor that is not conducting is said to be *off*, a transistor with an *unknown* on its gate is said to be *unknown*. We treat the source and drain of a transistor interchangeably.

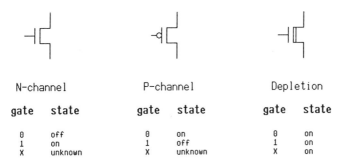

N-channel		P-channel		Depletion	
gate	state	gate	state	gate	state
0	off	0	on	0	on
1	on	1	off	1	on
X	unknown	X	unknown	X	on

Unfortunately, most MOS circuits have situations where a simple binary on/off model won't work. In the NMOS inverter that follows, if the signal IN is 0, then the depletion transistor connects OUT to VDD (power supply 1). If IN is 1, then the n-channel enhancement transistor connects OUT to VSS (power supply 0) as well. The resulting value of the output is 0 even though OUT is connected to both VDD and VSS because the enhancement transistor has lower resistance than the depletion transistor.

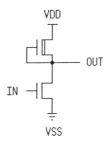

A node in the circuit can store charge, retaining its value when there is no path to a power supply. For example, in the following circuit, when the input A is 1, node B is connected to VSS. Node D remains at its value, 1 until C goes to 1, connecting it to VSS.

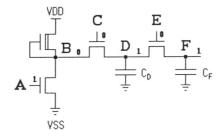

Let us assume now that C goes back to 0 and that node E is 0. Node D is not connected to any power supply, but it still retains its new value of 0. Node D is said to be *charged* to 0.

If node E now goes to 1, nodes D and F are connected. Assuming node F is charged to 1, the value of both nodes D and F will be 1, 0 or X depending on the capacitance of the nodes. If C_F is much larger than C_D, both will become 1. If C_D is much larger, both will be 0. If the two capacitances are approximately the same size, the result is undecided, X. This effect is called *charge sharing*.

Transistors

We model a transistor as a resistor. The gate of the transistor determines the resistance between its source and drain. A transistor that is *off* has infinite resistance and a transistor that is *on* has resistance R_{ds}. A transistor with an *unknown* gate has a resistance somewhere in between, which we can express as an interval of possible resistances $[R_{ds},\infty]$. The range of possible resistances on a transistor leads to a range of possible voltages on a node driven through that transistor. The node will become unknown if the entire range of voltages is not high enough to be a 1 or low enough to be a 0.

The resistance (R_{ds}) of any transistor can be approximated by a product of the intrinsic resistance of the transistor channel (R_{nchan}, R_{pchan}, or R_{dep}) divided by the Z-ratio, the transistor width/length for the individual transistor. The intrinsic resistances are different for each kind of transistor and depend on manufacturing details. The R_{ds} for an individual transistor can be computed once and stored with the transistor.

The simulator computes both the final value of a node and the time at which that value becomes valid. We estimate the delay with a simple $t = RC$ model. R is the resistance of the path to the power supply and C is the capacitance of the nodes that must be driven through that resistance. To improve the accuracy of both computations, we keep with a transistor three different resistances for R_{ds}: R_{static}, for use when calculating the final value on a node; R_{dynlow}, used when calculating the time for a node to go low; and $R_{dynhigh}$, used when calculating the time for a node to go high. The three different resistances let us obtain an accurate simulation in both resulting values and in delays. The difficulty is that we must derive the three different resistances to calibrate the transistor. Because manufacturing technology changes quickly, we keep the transistor calibration resistances in a *technology file*.

Simulation

The actual resistance of a transistor is a function of the voltage on its gate, source and drain. For simplicity, we assume that when the transistor turns *on*, the gate voltage changes very quickly. Then, as the transistor changes, the voltage on the source and drain change. The resistance curve for the transistor is different when both source and drain go to VDD than it is when they go to VSS. R_{dynlow} and $R_{dynhigh}$ are chosen to reflect that difference. R_{static} for a transistor is the resistance exhibited by the transistor when it is fully turned *on* and has VSS and VDD on its source and drain. This is the maximum resistance exhibited by the transistor, so it gives a conservative result.

```
class transistorClass(
    bits flags;
    integer transistorType;         # p- n- or depletion
    real rStatic, rDynHigh, rDynLow;  # resistances
    pointer(nodeClass) gate;
    pointer(nodeClass) source,drain;  # treated identically
);
```

Nodes and Stages

A node has a capacitance and connections to transistors. The value on a node may be *forced*, the value is set by the user and cannot be changed by the simulation; it can be *driven*, the node is connected by transistors to a forced node; or it can be *charged*, the node is not connected to a power supply, but is retaining its last value. Each node has its capacitance to substrate and separate lists pointing to the transistors whose gates are connected to the node and those whose sources or drains are connected to the node.

```
class nodeClass(
    integer value;
    bits flags;      # forced, watched, ...
    real capacitance;
    pointer(list) transistorGates, transistorNodes;
);
```

We decompose the transistor network into stages. A *stage* is a collection of transistors and nodes. Two nodes are in the same stage if they are the source and drain of a transistor that is *on* or *unknown*. All nodes in a stage change state together, since they are connected (separated by a closed transistor) or possibly-connected (separated by an unknown transistor). The stages of a circuit change as the simulation proceeds, so we dynamically recompute gate stages during the node evaluation as we simulate. For example, referring to the figure on page 251 all nodes are in separate stages. When node C is 1, nodes B and D are in the same stage. Although that stage is connected to node VSS, VSS is not considered part of the stage. Forced inputs to the simulation are not included in stages.

The Simulation Algorithm

The simulator is event-driven, as we saw with the logic simulator, though the evaluation of a single event is complicated by the transistor model. This section gives an overview of the transistor simulator, with the details explained in subsequent sections.

The basic simulation algorithm is a loop. Each iteration begins by removing the most recent event from the event queue and setting the value of the node to the new value. For each transistor that has the changed node on its gate, we perform *charge sharing* between the nodes on its source and drain. Then for all source and drain nodes of those transistors, we determine the *final values* by checking for connections to VDD and VSS.

We make the assumption that charge sharing happens immediately, while the transition to a final value takes some amount of time to complete. The final value of a node is determined by the connections from that node to forced nodes. VDD is always forced to 1 and VSS is always forced to 0. Users may force values of other nodes to 0, 1 or X. To find the value that a driven node will eventually reach, we compute the total resistance to 0 and to 1 and select the final value of the node if one dominates. The time that the new value arrives at the node is the product of the resistance to the power supply that drives the node times the capacitance on the node that must overcome to set the new value.

It is insufficient to merely handle a new event when a user sets a node to a new value, because if the node is the source or drain of a transistor, the value on that node may propagate through *on* transistors and *unknown* transistors to change other nodes. We propagate a user input through source-drains connections of *on* transistors.

```
procedure handleEvent(pointer(eventClass) e);
begin
  e.node.value := e.value;
  # charge sharing
  for all transistors (t) in e.node.transistorgates do
    if e.value sets t on or unknown then
```

Simulation

```
      chargeShare(t.source);
  # drive calculation
  for all transistors (t) in e.node.transistorgates do
      t.source.ready := t.drain.ready := true;
  for all transistors (t) in e.node.transistorgates do
  begin
      if t.source.ready then finalValue(t.source);
      if t.drain.ready then finalValue(t.drain);
      end;
end;
```

The boolean **ready** on a node indicates that we must determine the final value for the node. If two nodes are in the same stage, the final value calculation will compute their values at the same time. When we set the final value, we reset **ready** so we do not re-calculate the values of the nodes in a stage.

Charge Sharing

The idea of charge sharing is to accumulate all the capacitance charged to 0, 1 and X on nodes in the stage and from those charged capacitances, determine the resulting state of all the nodes. We traverse a graph-like data structure through *on* and *unknown* transistors, totalling the capacitances the nodes. If the total capacitance charged to 0 or 1 dominates, we set the node to that value, otherwise we set the node to X.

```
procedure chargeShare(pointer(nodeClass) firstNode);
begin
  putInList(stageList, firstNode);
  currentNode := first(stageList);
  c1 := c0 := cx := 0;
  while currentNode <> nullpointer do begin
    currentNode.visited := true;
    if currentNode.value = 1 then
        c1 := c1 + currentNode.capacitance
    else if currentNode.value = 0 then
        c0 := c0 + currentNode.capacitance
    else cx := cx + currentNode.capacitance;
    for all closed or unknown transistors (t)
        with currentNode as source{drain} do begin
      if not t.drain{source}.visited then
        appendToList(stageList,t.drain{source});
      end;
    currentNode := next node in stageList;
  end;
```

```
  if (c1+cx)/(c0+c1+cx) < vlow then result := 0
  else if c1/(c0+c1+cx) > vhigh then result := 1
  else result := X;
  for all nodes (n) in stageList do begin
    if result <> n.value then
      scheduleNewEvent(n,result,currentTime);
    n.visited := false;
    n.ready := false;
  end;
end;
```
vlow and vhigh are parameters that we use to set the limits on charge sharing -- how much one charge has to overwhelm the other to force the other nodes to its value. Because we want to be conservative, capacitance charged to X counts against a result of 0 or 1.

This evaluation ignores the possibility that there may be a path from the nodes in the stage to VDD or VSS. If there is such a path, it will be detected in the second phase of the evaluation and a new event will be scheduled to change the node to the power supply value. If this happens, this charge sharing calculation has not been superfluous, since charge sharing happens immediately and the driving by a power supply is scheduled in the future. The result is that the driven node immediately takes on its charge shared value, then changes to its driven value.

Evaluating Driven Stages

We evaluate each node that may be affected by the change in the netlist to determine its new value. We have two goals: to find the final value of the node and to determine the time at which that value is reached.

We assume that there may be more than one transistor pulling a node to VDD or VSS, so at each node we search for paths to VDD and VSS. We calculate an equivalent resistance for all paths to VDD and VSS, and calculate the resulting value and time from those resistances. The result of the resistance calculation is a *resistor divider*, represented as a pair of resistances, RH pulling the node high and RL pulling the node low. We use the magnitude of the resulting resistances to determine the final value of the node.

Simulation

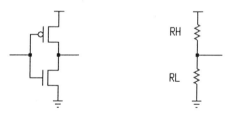

Unknown transistor values give a range of resistances, so all calculations to determine the resulting value on a node deal with interval values. Arithmetic operations on interval values are reasonably straightforward, the interval is always pessimistic. We represent an interval A as $[A_l, A_h]$.

$[A_l, A_h] + c = [A_l + c, A_h + c]$
$[A_l, A_h] + [B_l, B_h] = [A_l + B_l, A_h + B_h]$
$[A_l, A_h] - [B_l, B_h] = [A_l - B_h, A_h + B_h]$

We have two rules for combining resistances of nodes. A resistor divider with a series resistor can be turned into a new resistor divider. We encounter this structure when calculating the resistance of a node when we have already determined the resistances on the other side of a transistor.

RULE 1:

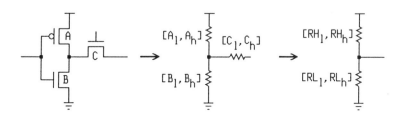

We can estimate the resistances with a clever approximation derived by Terman (1983). The approximation gives a reasonable approximation, but does not perform well if C is much larger than A and B. In MOS circuits, such situations are not common.

$RH_l = A_l + C_l + C_l A_l / B_l$ $\qquad RL_l = B_l + C_l + C_l B_l / A_l$

$RH_h = A_h + C_l A_h / A_l + C_l A_l / B_l$ $\qquad RL_h = B_h + C_l B_h / B_l + C_l B_l / A_l$

RULE 2:

Resistor dividers in parallel are added in parallel:

$RH_l = A_l \| C_l$
$RH_h = A_h \| C_h$

$RL_l = B_l \| D_l$
$RL_h = B_h \| D_h$

where $X \| Y = 1/(1/X + 1/Y)$

Knowing a resistance to VDD (RH) and to VSS (RL), we can compute the resulting *Thevenin* voltage that would result on the node: $V_{thev} = RL/(RH+RL)$. Since we are dealing with intervals, we compute the worst-case interval, $V_{thev} = [V_l, V_h] = [RL_l/(RL_l+RH_h), RL_h/(RL_h+RH_l)]$. We set the node value to:

1. If $V_l > v_{high}$ or node.value = 1 and $RL_l = \infty$.
0. If $V_h < v_{low}$ or node.value = 0 and $RH_l = \infty$.
X. Otherwise.

The extra clause for setting to 0 or 1 prevents us from setting a node to unknown when it could not have changed.

If the result value is 0 or 1, the delay for the node to take its new value is $D = R_{eff}C_{eff}$. R_{eff} is the resistance of the path from the node to the power supply to which the node is being pulled. We compute three additional resistances while we calculate the resistances RH and RL. These resistances are $R_{drive:1}$, the resistance from the node to VDD ignoring paths through *unknown* transistors; $R_{drive:0}$, the resistance from the node to VSS ignoring paths through *unknown* transistors; and $R_{drive:x}$, the resistance from the node to all inputs that are different than the current value of the node including paths through *unknown* transistors. $R_{drive:0}$ uses R_{dynlow}, $R_{drive:1}$ uses $R_{dynhigh}$ and $R_{drive:x}$ uses $R_{dynhigh}$ for connections to VDD and R_{dynlow} for connections to VSS. These resistances are series/parallel calculations on single values, not ranges like RL and RH. Resistance calculations for paths to 0 and 1 ignore paths through *unknown* transistors. Thus the calculated resistance may be larger than the actual resistance, so the delay will be larger. The resistance calculation for an unknown result includes all resistances, so it is as low as possible. In both cases we err on the side of conservatism.

C_{eff} is the amount of capacitance in the stage that must change to the new value. As we saw with the charge sharing calculation, different nodes in the stage may have different values initially. We calculate three capacitance numbers for the stage: C_1, the capacitance of nodes charged to 1, C_0, the capacitance of nodes charged to 0, and C_X, the capacitance of nodes charged to X. For the sake of conservatism, we assume that all unknown capacitance is the at the wrong value. Thus our times will be as large as possible. If the final value of the node is 1, then C_{eff} is C_0+C_X, if the final value of the node is 0, then C_{eff} is C_1+C_X.

If the final value of the node is X, then we have a different situation. Instead of wanting the transition to unknown delayed as long as possible, conservatism dictates that signals should go unknown as soon as possible. R_{eff} and C_{eff} should be chosen to be as small as possible, but realistically, to avoid problems with unknowns propagating through the simulation too quickly. We already discussed R_{eff}. It is not necessary to charge all capacitance in the stage to change the value of the current node, so a reasonable lower bound on the capacitance to charge is the capacitance of the current node. Careful analysis of the stage would admit a more clever algorithm for a higher lower bound on C_{eff}.

```
procedure finalValue(pointer(nodeClass) n);
begin
  calculateCapacitances(n, C0, C1, Cx);
  for each node (n) in stage s do begin
    calculateResistances(n, RL, RH, R0, R1, Rx);
    calculateVthevInterval;
    result := if vthev > vhigh then 1 else
              if vthev < vlow then 0 else X;
    delay  := if result = 1 then R1*(C0+Cx) else
              if result = 0 then R0*(C1+Cx) else
              Rx*n.capacitance;
    scheduleNewEvent(n, result, currentTime+delay);
  end;
end;
```

The procedure `calculateCapacitances` is a straightforward enumeration of the nodes in the group as we did in the charge sharing calculation. `calculateResistances` performs a recursive depth-first tree search to calculate the resistances and combines them as it works its way up the tree. It calculates the two resistance intervals RL and RH, to VSS and VDD. The intervals indicate connections only through *on* transistors as their high end, and through *on* and *unknown* transistors as their low end.

R0, R1 and Rx are the resistance paths to 0, 1 and X forced nodes used for the delay calculation (so they use the dynamic transistor resistances). R0 and R1 only include connections through *on* transistors, so they will produce a change to their value as late as possible. Rx includes connections through *unknown* transistors, so a transition to unknown happens as quickly as possible. In the following code, RL and RH are the resistance intervals to 0 and 1.

```
procedure calculateResistances(
  pointer(node) n;
  produces pointer(interval) RL, RH;
  produces real R0, R1, Rx;
);
begin
  n.visited := true;
  if n is forced 0 then begin
    RH := R1 := infinity;  RL := R0 := Rx := 0;
  end else if n is forced 1 then begin
    RH := R1 := Rx := 0;  RL := R0 := infinity;
  end else begin
    for each on transistor (t) with
          source{drain} connected to n do begin
      if not drain{source}.visited then begin
        calculateResistances(drain{source},
          lRL, lRH, lR0, lR1, lRx);
        combine R(t) with lRL and lRH using rule 1
        combine lRL and lRH with RL and RH using rule 2
        if t is not unknown then begin
          R1 := parallelResistances(R1,(t.dynHigh+lR1));
          R0 := parallelResistances(R0,(t.dynLow+lR0));
        end;
        Rx := parallelResistances(Rx,
          (t.dynHigh min t.dynLow)+lRx));
      end;
    end;
  end;
end;
```

Managing the Event Queue

We have seen that we have two kinds of events changing node values: *charge sharing events* and *final value events*. During simulation, we make the assumption that when we compute a new value for a node, the new value represents a more accurate description of the circuit than any old values in the event queue. Therefore, when we add a new charge-sharing event, we remove all pending events for that node from the event queue. Any old charge sharing event must be superseded by the current one, and the current charge-sharing event is followed by a final-value computation that will supersede the old final-value. We add the new charge sharing event only if its value is different from the node's current value.

When we add a new final-value event, we would remove any old final-value event, but since each final-value calculation is preceded by a charge-sharing calculation, the new charge-sharing event has already removed the old final-value event. We only add the new event if the final-value value is different from the current node value or the charge-sharing event's value if there is one. There may be at most two events in the event queue for each node, a charge sharing event and a final-value event.

Transistor Simulation Example

In this section, we follow the simulation of a sample circuit. The circuit is an implementation of a CMOS exclusive-OR gate, shown in the following figure. The circuit has an inverter on the left to invert the a input, and a complicated structure on the right that acts like a transmission gate for b when a is 0 and acts like an inverter for b when a is 1, with a and nota acting like VDD and VSS.

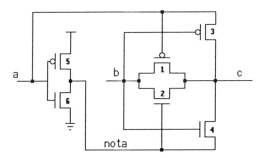

We start the simulation by setting a = b = 0. a turns transistor 5 *on* and transistor 6 *off*, so nota is 1. Since a is 0 and nota is 1, transistors 1 and 2 are *on* so c = b = 0. Transistors 1, 2, 3, and 5 are *on*. Inputs a and b are forced by the simulation, so they can't change.

First we change input a from 0 to 1.

Time 1. Event: a→1.

Since a is a user input, we propagate the change through transistor 3 which is *on*, setting c to 1 also. We note that we must recompute the final value for node c. Since transistors 5 and 6 changed, we must re-evaluate the nodes they touch. Node nota is the only node that is not a power supply. There is no charge sharing from power supplies, so we calculate a new final-value event for node nota to go to 0 at some time in the near future. The time is short since the wiring capacitance on nota is low.

Examining node c, we find it is in a stage with nodes a and b since transistors 2 and 3 are *on* (nota is still 1). Again there is no charge sharing, and since c is driven low from b and high from a, it must go to X, since we cannot resolve the conflict. We schedule the transition for c in the future also.
Node values: a=1, b=0, nota=1, c=1. Event queue: nota→0, c→X.

Time 2. Event nota→0.

nota turns *off* transistor 2, so we re-compute node c. Now the stage contains only c and a. We remove the old transition for node c, even though there is no charge sharing. Node c is driven only by node a through the transistor 3. We schedule the event for node c to go to 1, but since it is already 1, there is no new event.
Node values: a=1, b=0, nota=0, c=1. Event queue: empty.

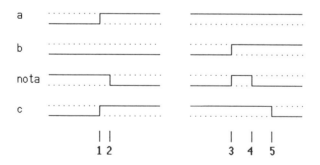

Now we change input b from 0 to 1.

Time 3. Event b→1.

Since transistors 1 and 2 are both *off*, the value of b does not propagate to any other nodes, but the new value of b does turn transistor 3 *off* and transistor 4 *on*. We must recompute the value of nodes c and nota. They are in the same stage, so first we perform the charge sharing calculation.

Let us assume that the capacitance on node c is much larger than that on node nota. We generate a charge sharing event to change nota to 1, the current value of c. We also evaluate the two nodes to determine their final values. They are both driven by VSS through transistor 6. The resistance to node nota is less than the resistance to c, so the transition for node nota is earlier than the transition for c.
Node values: a=1, b=1, nota=0, c=1. Event queue: nota→1, nota→0, c→0.

Time 3. Event nota→1.

This is the charge-sharing event that was scheduled immediately. Since nota is 1, transistor 2 turns *on*. Nodes b, c and nota are now all in the same stage, since transistors 2, 4 and 6 are *on*. Node c and nota are at the same value, so there is no charge sharing. Now we determine the final values for nodes nota and c. Node c is being pulled low through transistors 4 and 6. It is also being pulled high through transistor 2 to node b. Assuming the resistances are comparable, we must schedule a final value event for node c to go *unknown*. Node nota is also connected to c, so we might schedule its final event to be unknown also, but it is connected to 0 through one transistor, 6, and to 1 through two transistors in series, 2 and 4. The two transistors have a higher total resistance, but they also have higher individual resistance, since they are n-channel transistors that do not pass the high value well. Therefore, the final value for nota should be 0 because the resistor divider shows a much lower resistance path to VSS. Of course, the previous final value events for nota and c are discarded.
Node values: a=1, b=1, nota=1, c=1. Event queue: nota→0, c→X.

Time 4. Event nota→0.

Now nota turns *off* transistor 2. Node c and nota are still in the same stage, but there is now only one path to a power supply, through transistors 4 and 6 to VSS. We remove the old event for node c and schedule the final value of 0.
Node values: a=1, b=1, nota=0, c=1. Event queue: c→0.

Time 5. Event c→0.

Node c reaches its final value.
Node values: a=1, b=1, nota=0, c=0. Event queue: empty.

Although no signals were set unknown, there were events in the queue to set c unknown. Those events were subsequently replaced by transitions to a known values. If there were a very low capacitance on node c, it might go unknown for a short time. We saw a glitch in the waveform for nota caused by charge sharing. Consider the effect of making transistor 6 a high-resistance transistor, or consider the effect of making transistors 5 and 6 into a large NAND gate. Then the resistance for node nota through transistor 6 could be comparable to the resistance through transistors 2 and 4. The result would be that node nota would go to unknown at time 4 after the charge sharing forced it high. Node c would go unknown and would stay unknown. Even though the circuit in reality would eventually resolve to a value, probably the correct one, the delay could be large. The circuit represents a potential problem, so it should be flagged. Changing its output to unknown would serve to identify the potential problem.

Debugging Commands

As we discussed during our description of a logic simulator, we need a method to view the nodes before and after the current node. The concept is the same with transistors as it was with gates. show instance lists information about a transistor, show node lists the node value and its capacitance, show after lists information on transistors of which the node is the gate and show before lists information on transistors of which the node is source or drain. The transistor information includes the nodes on its gate, source and drain, the transistor type, width and length and its state.

Enhancements to the Transistor Simulation Algorithms

Scheduling Charge Sharing Events

When we do charge sharing, we assume that charge sharing is unwanted, so our conservative attitude leads us to schedule the new event immediately. A more precise calculation would take into account the capacitance of the nodes that are changing and the resistance of the transistor that is connecting them. The delay would be another RC calculation, with capacitance accumulated according to charged value.

Tighter Bounds On Charge Sharing Results

The charge sharing evaluation assumes that all unknown transistors are closed (conducting), but this may lead to optimistic results. A node charged to 1 but connected through an *unknown* transistor should not contribute to the result becoming 1. A resolution of this problem requires that the capacitances for each node in a stage be calculated as a range. The low end of the range is the accumulation of all capacitance that is reached only through *on* transistors, the high end of the range includes capacitance through *unknown* transistors. We set the result to 0 only if the least capacitance at 0 dominates the maximum 1 and X capacitance, and to 1 only if the least high capacitance dominates the maximum 0 and X capacitance. The resulting more conservative calculation of the result value is:

```
if (c1.h+cx.h)/(c0.l+c1.h+cx.h) < vlow then result := 0
else if c1.l/(c0.h+c1.l+cx.h) > vhigh then result := 1
else result := X;
```

The form of the calculation is the same as our simpler charge sharing calculation. The highest possible value for the node results when connections through unknown transistors to 1 and x are *on* and connections through *unknown* transistors to 0 are *off*. Thus, when checking for a low value, we use c1.h, cx.h and c0.l. The converse holds when checking for a high value. A more detailed examination of the topology of the network could indicate a connection through an *unknown* transistor to both a 1 region and a 0 region, from which we could get tighter bounds.

The major difficulty with this range calculation is that it requires a separate calculation of c0, c1 and cx for each node in the stage, since each node will "see" different amounts capacitances at 0, 1 and X. Each calculation requires a recursive search through the stage. However, much of the added calculation can be absorbed if the charge sharing calculation is merged with the final value computation, since during the final value computation, we already do a recursive search through the tree for each node in the stage.

Better Accuracy in Gate-Like Structures

Merging the two computations also has an advantage that we can run the event queue more accurately. For example, suppose we have a NOR gate with both inputs 0, as shown in the following figure. The output is 1. When one input changes, the simulator assumes that the voltage drops according to the first curve and schedules the event to change the value of OUT accordingly. In reality, when the second input changes, there is now less resistance to VSS so the voltage drops more quickly. However, the simulator sees that the node is still high so it assumes the higher, more conservative curve and replaces the final value event with a new event.

Although the delay of the new event is less, the resulting event falls at a later time because it was scheduled later. Since we do the charge sharing first, the first final value event is gone from the queue when we calculate the second one. An improvement suggested by Schaefer (1985) is to check the first final value event when we schedule the second one, taking into account that the node was already approaching the desired value and schedule the new event accordingly.

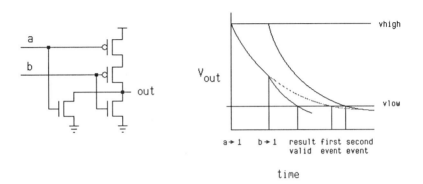

Improved Accuracy of Transistor Resistances

The resistance of a transistor is not a simple product of a basic resistance and the length/width of the transistor. For a more accurate simulation, we take into account the *context* of the transistor. An enhancement transistor can be used as a pulldown device, a pullup device, a bootstrap device or a pass transistor. These special uses can be determined with a local check of connections in the netlist. For an accurate resistance calculation, we can use circuit simulation to derive accurate static and dynamic intrinsic resistances for transistors in different contexts with different sizes.

Charge Decay

We noted at the beginning of the discussion of transistor simulation that nodes retain their values when they are disconnected from the power supply. In actuality, the charge on a node leaks away over time. The time is typically in milliseconds, very large by integrated circuit standards, and most such signals are dynamically refreshed frequently. In our simulation, when there is infinite resistance to all power supplies, we can schedule a charge leakage event so that a node that is left charged to a value will eventually become unknown after a suitably long time. This leakage time should be much shorter than the real decay time since a single run of the simulator would not simulate sufficient duration of time, but a user can set it to be longer than his expected refresh time to catch nodes that he is not properly refreshing. Unfortunately, the event queue does not handle large numbers of distant events efficiently.

More Accurate Signal Values

A major advantage of the simulator, unknowns, is also a major cause of problems. Since we try to be conservative in the simulation, we may generate unknowns unnecessarily or unknowns may remain in the simulation when in reality they would have been resolved. For example, in the Schmitt trigger circuit that follows, assume IN is 1 so OUT is 0. When IN goes to 0, OUT goes to X because the resistance of the path through transistors A and B is finite and the resistance through transistor E is also finite. Since node OUT changed, it is scheduled to be re-evaluated. When it is re-evaluated, though, we must assume that transistor E is still conducting, since its gate is *unknown*. The dotted line feedback path keeps OUT unknown. It can never go to 1. If IN changes back to 1, OUT remains unknown because of the lower feedback path.

In reality when IN goes to 0, the voltage on OUT drops to some intermediate value so transistor E begins to turn off. The higher resistance of transistor E allows OUT to rise a little farther, so E turns off a little more. Eventually, E turns off fully and OUT is 1.

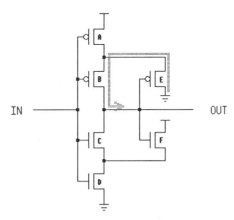

The difficulty here is caused by the three-state model we have chosen to represent signals. We use the X state as an error condition (interval [0,1]) and as an intermediate value (interval [vlow,vhigh]). We could allow more intermediate values or represent signals as voltages and let transistor resistances vary with the voltage on the gate (Schaefer 1985). We would then let events change voltages over time. The transistor evaluation would then be more time-consuming but more accurate, and circuits like the Schmitt trigger would simulate correctly.

Multiply-Connected Paths

Our resistance calculation does not recognize cycles. If we design a circuit with a cycle, the simulator will only keep the first branch it finds, as shown in the top pair of figures, below. The result is an overestimate of the resistance of the circuit. This overestimate leads to greater delays, so it meets our criterion of conservatism, but the delays may be much longer than the real delays for the circuit. A common cyclic circuit is the CMOS transmission gate, the lower pair of figures, below. It is possible to detect transmission gates in the netlist and model them as a single transistor in the simulation. Alternatively, we can order transistors in the node lists in order of increasing resistance, so the least resistance transistor will be used. However, such **solutions do not handle the general case. A more detailed** analysis of such cycles could solve the problem at the cost of longer execution time, of course.

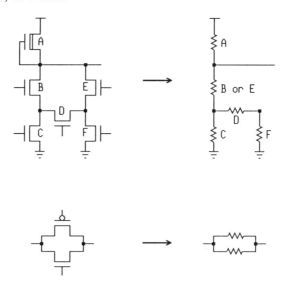

Performance Improvements

We simulate circuits as collections of transistors rather than as gates because the transistors can be bidirectional. However, most transistors in an integrated circuit are not bidirectional, they make up gates. We can take advantage of simple cases by combining transistor stacks into single "compound transistors" for simulation, like the circled transistors in the following figure. Another possibility is to recognize transistor configurations that make up entire logic gates and to treat them as single units. Of course the simulation accuracy suffers slightly, but the performance gain can be enormous.

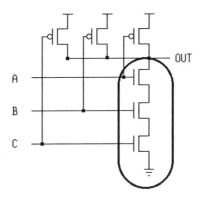

Static Checks

The netlist representation of a circuit that we use for simulation is also valuable for checking errors in circuit design without simulation. The errors we can find include questionable use of transistors that may indicate layout errors as well as constructs that are circuit design problems. In this section we discuss some of these checks that have been found to be valuable to designers.

Every Node Can Be Pulled Up and Down

For every node in the circuit, there should be paths that lead to VDD and VSS or a path that leads to an external input. If there is no such path to a power supply or input, then this node cannot ever reach that power supply value. This usually indicates an error in circuit design.

Check Transistor Sizes Versus Node Capacitance

If a small transistor (high resistance) drives a large capacitance, the delay to drive that node will be large. This check calculates the RC time to drive each node high or low and flags any time that exceeds some limit. Another check on transistor sizes compares the resistance of the pulldown path to the resistance of the pullup path for all nodes. If the resistances differ by too much, the node is flagged as an error.

Both these checks indicate possible circuit design problems, but both can flag structures that are correct. A user may not care how slowly some nodes change. There are usually limits, though, so the critical figure of merit cutoff number should be something that users can change.

VDD and VSS Separated by One Transistor

If one transistor separates VDD and VSS, then if that transistor ever turns *on*, there will be a short circuit between those nodes. Any such transistor should be flagged as an error. This structure usually indicates a missing transistor or a node accidentally shorted to VDD or VSS.

A true short between VDD and VSS must be caught during circuit extraction, if at all, since the extractor will merge the node names, eliminating one of them. A static check could determine if both nodes were present. If not, they were probably shorted together. In addition, if the two names were aliases for the same node, they were shorted together.

NMOS Signal With a Threshold Drop Gating a Pass Transistor

An NMOS pass transistor passes a signal that reaches a maximum of one transistor threshold drop below the high voltage power supply. If that signal is used to gate another pass transistor, the resulting signal through the second pass transistor is two threshold drops down. The resulting signal may be too low to turn on a n-channel transistor, so the circuit design is probably in error.

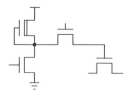

Non-standard Transistor Usage

Different kinds of transistors have typical uses in circuits. N-channel transistors pull nodes to VSS, p-channel and NMOS depletion transistors pull nodes to VDD. An n-channel transistor with source tied to VDD or a p-channel transistor with source tied to VSS is probably in error.

In NMOS, a depletion transistor without drain tied to VDD is usually an error. The gate of a depletion transistor is usually connected to the output node, or if it is not, it should be part of a *superbuffer* structure. Both these cases can be tested with local checks in the circuit.

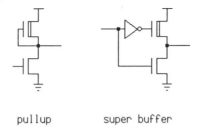

pullup   super buffer

Self-gated transistors are those that have their gate connected to source or drain. Although this structure is typical for NMOS depletion transistors, it is unusual for other transistors.

A Node That Doesn't Drive Anything

Another easily-checked problem is a node that doesn't drive anything. An unconnected node probably is an output of the circuit. Since users typically watch outputs, it is not unreasonable to require that outputs be declared. The result is that "undeclared outputs" are nodes in the circuit that do not drive anything. These nodes are probably the result of a circuit design error.

List Capacitors

A capacitor is made by shorting together the source and drain of a transistor. They are rarely made intentionally. A command to list all such transistors will flag possible errors.

Timing Analyzer

A *timing analyzer* is a program that analyzes the timing of a circuit without regard to the actual signal values in that circuit. We enumerate paths through the circuit. We use the gate delays or transistor delays to find the worst-case paths and report them to the user. A timing analyzer gives results much faster than simulation, but it may be less accurate. Because it does not have signal transitions, it must use worst-case transitions at all times to be conservative. The result may be over-conservative to the point of unusability. Further, a timing analyzer may report paths as *critical* that are never taken in operation. These problems are addressed by more accurate timing analysis and selective timing analysis. A detailed discussion is beyond the scope of this chapter, but you may pursue descriptions by Ousterhout (1983) and Jouppi (1983) in the references.

Network Comparison

A *network comparison program* compares two netlists and reports differences (Ebeling and Zajicek, 1983). It can be used to check that the netlist extracted from the layout matches the netlist entered during circuit design. Network comparison is a *graph isomorphism* problem. The program attempts to identify equivalent nodes and instances in the two netlists. The comparison fails if the two circuits do not match, it can also be directed to fail if expected capacitances on nodes are not within allowed tolerances.

Multi-Level Simulation

All the types of simulation we have examined -- behavioral, logic and transistor -- include timing and an event queue. It is possible to perform all three kinds of simulation in one program that shares the same event queue. A *multi-level* or *mixed-mode* simulation allows a user to simulate a part of the circuit, for example an ALU, as transistors to gain accuracy, simulate other parts of a microprocessor as logic gates and simulate memory as a functional model. The resulting simulation would be both fast and accurate.

The changes required to merge the different types of simulation are surprisingly minor. When we remove an event from the event queue and change the node value, we check charge sharing and final values for nodes connected to transistors as we did in the transistor simulation, but we also check inputs to functional models and logic gates and evaluate them. All results go back on the one event queue. The biggest change is in the netlist file, which must accommodate all three kinds of data.

This method works well as long as the functional models accept and produce only the ternary 0, 1 and X that the logic and transistor simulations require. In order to make this work properly, we would have to include some code around functional models to assemble bits into numbers when needed and to break vectors back into bits. This translation is most conveniently done by users as part of the functional model interface. This way functional models communicate using single-bit values through the event queue, so no changes are needed in the event queue.

However, functional models tend to change many bits at once, filling the event queue with multiple events. Since the bulk of the time in the simulator is spent manipulating the event queue, we would like a more efficient method. An alternative method is to add another kind of event to the event queue, a multiple-node event, that contains the multiple-valued output from a functional model, represented as the functional model signal name, the single-bit nodes to which it can be translated, and the value on the signal. When the event is removed from the event queue, if it goes to another functional model, it is transferred without translation, if parts of the signal go to transistors or to logic gates, the number is decoded and the individual bits transferred.

Parameterized Models

Fixed functional models can be implemented as a single procedure or module that is wakened by the event handler and adds events to the event queue. Parameterized models face the same problems with parameterized connections that we saw with parameterized layout. The model must have two parts, one to initialize the cell and another to run in the simulation. The initialization part sets the number of connectors and their names, the simulation part interacts with the simulation. Since the numbers of inputs and outputs may change, the size of the inputs and outputs lists in the eval procedure is set at simulation time.

High-level Input

We have already discussed commands to provide input to the simulator. We can set any node to high, low or unknown. However, it can be very tedious to provide input one bit at a time for a bus that is sixteen bits wide. Therefore, many simulators provide a facility to define busses and to set the values of all nodes in a bus with one command, converting hexadecimal, decimal, octal or binary to individual signal values:

```
vector a a0 a1 a2 a3 a4 a5 a6 a7
set a 14
```

The complexity of stimulus for a simulator is not limited to vectors. We may wish to set the values of many signals, or conditionally set their values depending on the current simulation time or on the state of internal variables in the simulation. We can add these loops and conditionals to the simulator input language giving a *stimulus language*. The stimulus language has access to all internal variables and can set the values of nodes immediately or at some time in the future, just like we can as users at a terminal.

Users can build procedures in the stimulus language that enable them to exercise a simulation with higher-level commands (Van Egmond 1984). For example, for a simulation of a processor, a user may write a single command to increment a register. Rather than having to specify the register number, the ALU operation and a number of cycles to run, he gives one command to set all signals and step the clocks.

One powerful use of a stimulus language is to develop a set of *tests* for a circuit. A test includes not only the stimulus for the circuit, but also the expected response from the circuit. Thus, it supplies the full verification that the circuit functions as desired. The set of responses may be captured from a run of the simulation or provided by the user. A stimulus language with test facilities is called a *test language*. We can subsequently translate the test language into *test vectors* for an IC tester. Test vectors consist of both the inputs to and expected outputs from a correctly operating circuit.

Testing Integrated Circuits

Overview

There are two kinds of tests performed on integrated circuit: functional tests, to ensure that the circuit being built is the circuit a user intended to build; and production tests, to ensure that each chip has been manufactured correctly. The "testing" problem in integrated circuits refers to production test.

During the integrated circuit manufacturing process, a single chip may be bad due to a number of causes, such as a particle settling on it during the manufacturing process. When the chips are completed, all parts are tested to ensure that they were made correctly. A small defect will cause only a small part of the chip to be in error. A production test exercises the chip to discover parts of the chip that are defective. We require a set of test vectors that allows us to detect such tests.

One of the design tasks is the job of developing the test vectors used for that testing process. In the preceeding section, we touched on the idea of a test language to describe the tests. In this section, we introduce testing concepts and describe a method for verifying that the tests we developed exercise the chip sufficiently that we expect to catch errors in manufacturing.

Types of Faults

During manufacture, a defect typically results in part of a layer missing, or a piece of a layer existing where it should not. A defect may or may not be fatal to the operation of the circuit. In the following figure, fault 1 would add a little capacitance to the wire, fault 2 would probably break the metal wire, but might not affect the operation of the inverter, instead it might shorten the lifetime of the part. Fault 3 would surely be fatal since the contact is missing. In general, fatal defects result in wires being cut or shorted and transistors being left permanently open or closed. Many attempts have been made to test for the kinds of faults such defects would cause. However, because nodes in an integrated circuit can store charge and because there are so many wires, test software uses a simpler model of integrated circuit defects.

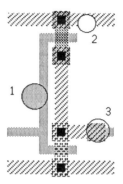

The most common model of faults in integrated circuits is *stuck-at fault*. Stuck-at faults are faults that result in a node or input of a gate never being able to change value. Nodes are stuck-at 1 (SA1) or stuck-at 0 (SA0). Many real defects on chips result in nodes being stuck, and many non-stuck at faults will be caught with a test set that was designed to catch stuck-at faults, but the real relationship of genuine defects to the stuck-at fault model is unknown.

Fault Simulation

In this section we describe methods for *fault simulation*. Fault simulation simulates a circuit as if it had faults in it and reports the fraction of faults that were detected by the test vectors the user supplied. The user may then add more test vectors to the test set to find faults that were not detected initially.

A fault simulator does not generate tests automatically, it merely gives an indication of the quality of the test a user has developed. Although software exists for automatically generating test vectors, such software usually requires substantial restriction in the design methodology and generates an unacceptably-large set of test vectors. A large number of test vectors requires a large amount of time to test the part. Fault simulation plays an integral part in attempts to automatically generate good quality tests.

Overview

The idea of fault simulation is to force a node to a stuck-at value creating a faulty circuit, then run the simulator, usually a logic simulator, using the test vectors with the faulty circuit and with the good circuit simultaneously. We keep track in the simulator of the differences in node values in the two simulations. If the differences reach an output, then the simulation *detected* the error. If the test set ends and the simulation has not detected the fault, then that fault is reported as an undetected fault. The results of fault simulation are the fraction of detected faults, called the *fault coverage*, and the list of undetected faults. This process is also called *fault grading*.

For example, in the following figure, suppose we wish to test node d for being stuck at 1. If our test vector inputs are: abc = 110, 001, we can't detect the fault, since the value at the output node, e, is indistinguishable from the correct value for both vectors. For the first vector, the output of the AND gate will be 1 anyway, so we can't tell if the node is SA1 or driven to 1 correctly. To test the node, it must be *controlled* to be opposite the faulty value. In the second vector, node d will be 0 in the correct simulation, and SA1 in the faulty simulation, but since c is 1, the output node e will be 1 regardless of the value on node d. To detect the difference, we must be able to *observe* the value of the node as we control it to its value. We can catch the node with the vector 010 which controls the node to 0 in the correct case, and observes the value on the node simultaneously.

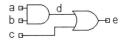

It is clearly unreasonable to run two simulations per node in the circuit to determine the fault coverage. Two techniques have been derived to speed up the simulation, *parallel fault simulation* and *concurrent fault simulation*.

In parallel fault simulation, instead of only two simulations, we perform many simulations, one on a fault-free circuit and many others each with a different fault added to the circuit. Each node, instead of having one value, has several, usually dictated by the word size of the machine. Logical operations are done on multiple simulations simultaneously. Parallel fault simulation speeds up the fault simulation process by a factor of the number of simultaneous simulations.

In concurrent fault simulation, each value in the simulation has associated with it a fault identifier. Nodes also have a fault identifier to indicate which fault corresponds to their stuck-at values. Each node can have a different value for each fault. However, nodes do not have large numbers of different values because the simulation constrains them to 0, 1 and X (or to speed up the simulation, just 0 and 1). As the simulation runs, faults from two different faults may result in the same value at a node. Those two fault values can be merged into a single value that represents the value for both faults. We are not concerned with the location of the fault, the fault identifier will provide that. We care only with the fault's presence or absence.

When a fault is detected, we remove all references to that fault number from the simulation. The simulation proceeds until the entire set of test vectors has been processed.

Performance Enhancements for Fault Simulation

Although concurrent fault simulation is significantly faster than other fault simulation techniques, it still takes an enormous amount of time to simulate every fault in a large circuit. Fault simulation can be made faster with a number of techniques. First, a simpler simulator can be used -- a logic simulator instead of a transistor simulator. Timing may be eliminated or simplified, unknown values may be omitted. Thus the simulation used for fault grading will be simpler than the simulator used for verification.

In addition, we may not simulate all the possible stuck-at faults in the circuit. The faults to be simulated may be set by a user, or they may be a randomly-selected fraction of the total faults possible in the circuit. Thus, the percentage fault coverage is a statistical expectation. A statistical result from fault simulation is not totally useless. The fault simulation process is inexact for other reasons as well. The faults we investigate in the circuit do not precisely model the effects of faults in integrated circuit manufacture and the entire simulation process is an approximation of the operation of the real circuit. The result is an approximation of the real fault coverage, which is useful for grading the test vectors despite its weaknesses.

Exercises

Programming Problems

1. Write an embedded language for behavioral simulation. Use it to simulate the microprocessor described in this chapter.

2. Write an embedded language netlist generator to generate input for a logic simulator.

3. Implement an event-driven logic simulator. Include the functions AND, OR, NAND, NOR and INVERT for three-valued logic (0, 1, X). Implement constant delays.

4. Write a procedure that simulates a PLA given a code file. Inputs and outputs may be 0, 1 or X.

Questions

1. Give a method for implementing events and inertial delays that does not require that we remove events from the event queue. What are the advantages and disadvantages of your method?

2. Write the truth table for an AND gate in a five state simulator where the signal values are 1, 0, X, R (rising) and F (falling). What changes would you need to make to the logic simulator gate evaluation to use all the states?

3. Draw a circuit using gates that settles immediately in reality, but would oscillate continually in a unit-delay logic simulator.

4. Write a BNF for a logic simulation netlist. The primitives in the language should include AND, OR, NAND, NOR and INVERT. Signals are 0, 1 and X. Include a method for referencing functional models and write down the semantics of the functional model reference.

5. What are the advantages and disadvantages of implementing a test language as an embedded language?

6. What data structures does a user need in a test language?

7. Draw a circuit composed of logic gates that has an *undetectable fault*, a fault that cannot be propagated to the outputs no matter what inputs are applied to the circuit.

References

Simulation

R. Bryant, "An Algorithm for MOS Logic Simulation", *Lambda Magazine*, Fourth Quarter, 1980.

R. Bryant, "A Switch Level Model and Simulator for MOS Digital Systems", *IEEE Transactions on Computers*, vol. C-33, No. 2, February, 1984, pp 160-177.

B.R. Chawla, H.K. Gummel and P. Kozah, "MOTIS -- An MOS Timing Simulator", *IEEE Transactions on Circuits and Systems*, vol. CAS-22, December 1975.

M. d'Abreu, "Gate-Level Simulation", *IEEE Design and Test*, December 1985.

T.M. Lin and C. Mead, "Signal Delay in General RC Networks", *IEEE Transactions on Computer Aided Design of Integrated Circuits and Systems*, vol. CAD-3, October 1984.

T.J. Schaefer, "A Transistor-Level Logic-with-Timing Simulator for MOS Circuits", *Proceedings of the 22nd Design Automation Conference*, 1985.

C.J. Terman, *Simulation Tools for Digital LSI Design*, PhD Thesis, MIT/LCS/TR304, Massachusetts Institute of Technology, 1983.

E. Ulrich and I. Suetsugu, "Techniques for Logic and Fault Simulation", *VLSI Systems Design*, October, 1986.

J.K. White, A. Sangiovanni-Vincentelli, *Relaxation Techniques for the Simulation of VLSI Circuits"*, Kluwer Academic Press, 1987.

Testing

K. Son, "Fault Simulation with the Parallel Value List Algorithm", *VLSI Systems Design*, December 1985, pp 36-43.

K. Van Egmond, "Software Unites Test Program Development with Circuit Design", *Electronic Design*, November 15, 1984.

General

A. Aho, J.E. Hopcroft and J.D. Ullman, *The Design and Analysis of Computer Algorithms*, Addison-Wesley, 1974.

C. Ebeling and O. Zajicek, "Validating VLSI Circuit Layout by Wirelist Comparison", *Proceedings of the 1983 International Conference on Computer-Aided Design*.

N.P. Jouppi, TV: An nMOS Timing Analyzer, *Third Caltech Conference on VLSI*, 1983.

J.K. Ousterhout, "Crystal, A Timing Analyzer for nMOS VLSI Circuits", *Third Caltech Conference on VLSI*, 1983.

INDEX

A

Aho, A.V.   17
alignment mark   **63**
arrogant software   137
artwork   *see* layout
Ayres, R.   143

B

Baird, H.S.   187
Baker, C.M.   230
Bastian, J.D.   225
Batali, J.   120, 137
behavioral description   6
behavioral language   6
behavioral simulator   6, **236-242**
Bentley, J.L.   92
bin
   for pen plot   76
   for raster plot   78
bit plane   57
bloat   *see* resizing layout
BNF
   definition of   **18**
   for layout file   18
   for netlist file   226
   for PLA code file   154
bounding box   20
Brayton, R.K.   153
Breuer, M.A.   170
Bryant, R.   250
Burstein, M.   176

C

Caesar   103-104
capacitance
   of transistors   5
   of wires   5
cell
   definition of   **12**
   in placement and routing   168
   instance in placement and routing   168
   separation of data from tool   32
   type of   13
cell compiler   141
cell manager   38
channel   168-169
   definition of   **169**
   density of   169
channel route   10, **174-176**
   channel in   169
   constraint graph in   175
   dogleg in   175
   track assignment in   175
   track in   169
charge sharing
   and unknowns   252
   as scheduled event   264
   event   260
   in transistor simulation   **255**
charge storage
   decay of   266
   in transistor circuits   251
charged node   252
Chawla, B.R.   250
Cho, Y.E.   112
circuit design   7
circuit extractor   *see* extractor
circuit simulator   7, 236

clipping   47
clock
    in a simulator   246
    phase of   248
CMOS   2
Cohen, D.   48
color   57
    equivalent stipples for   2
    on black and white display
        60
    opaque   58
    stippled   59
    transparent   57
color map   **43**
command driven   70
command file   70
compaction   110-112
compactor
    for symbolic layout   111
component
    in symbolic layout   109
concurrent fault simulation   277
connector
    in layout language   123
conservatism
    and false errors   184
    and unknown propagation
        267
    in computing resistance   258
    in DRC   184, 210
    in final value computation
        259
    in simulation   235
    in transistor resistance   253
constraint graph compaction
    111-112
contact   3
contact cut layer   2
controlability   277
Conway, L.   1, 39, 128
corner stitched tiles   95-97
cost function   172
    relation to layout quality   173
critical path
    in symbolic layout   111
current command mode   87

cursor
    in Caesar   103
    in dynamic display   63-64
    on a display   61
cut and paste   101
    in Caesar   **104**
cut count   170

D

d'Abreu, M.   242
data path element   159
database   107
datapath
    bit slice in   160
    definition of   159
    elements of   160
    floorplan of   159
    language specification of   164
datapath compiler   **159-167**
    bus connections in   161-162
    bus span in   163
    busses over cells in   162
    cell design for   167
    control line drivers in   166
    heavily-parameterized cells in
        161
    input to   163
    optimization of   162-163
    parameters to   165
    placement improvement in
        163
    power and ground connections
        in   166
    track assignment in   163
defects
    causes of   275
    examples of   276
delay
    cause of   **5**
demand paged virtual memory   38
density
    at a location   169
    of a channel   169
depletion transistor   250
design hierarchy   12
design restriction   14
design rule checker   *see* DRC

Index

design rules 184-185
   definition of 9, 183
   example of 185
Dholakia, S. 145, 162, 167
display output device 43
dither *see* stipple
dragging **64**
DRC 9, **185-221**
   acute angle errors in 216
   connectivity in 210
   corner ambiguity in 215
   data structure alternatives for 187
   deciding width or spacing in 206
   edge algorithm for **203-209**
   edge file in 187
   enclosure check in 207
   examining an edge in 205
   false errors in 184, 210, 214
   flat netlist requirement in 183
   flattening hierarchy in 187
   gate overlap in 208
   hierarchical 228
   ignoring perpendicular edges in 213
   NOCHECK in 228
   of edges **187**
   of objects 186
   OR rules in 209
   polygon identifiers in 210
   polygon merge in 211
   reporting errors in 213
   similarity with extractor 183
   stamping in 210
   technology file for 218-220
   vertical edges in 206
   with corner stitched tiles 97
   with objects 186
   with raster 230
   with resizing 201-203
   with trapezoids 230
Dunlop, A.E. 170, 174
Dutton, R.W. 225
dynamic display 62-65
   definition of **62**
   implementation of 64-65

E

Ebeling, C. 272
edge
   active 189
   boolean operations on 198
   border 192
   crucify 208
   definition of **188**
   direction of 188
   fast sorting 221
   horizon 204
   in DRC 188
   intersection of 191
   order problems 216
   processing algorithm 189-191
      example of 193-198
   sorting of 189
   stamping of 210
   vertical edge redundancy 188
edge file 187
Edgington, D. 145
editor state 85
electrical rules checker 10
Ellement, M. 225
embedded language **119**
ERC 10
event
   charge sharing 260
   final value 260
   in simulation 240
   user interface 248
event queue 248-249, 260-261
   definition of **241**
event-driven simulator **240**
extractor 9, **221-227**
   capacitance in 223
   DISCONNECT in 228
   electrical nodes in 222
   finding transistor connections in 223
   hierarchical 228
   horizon in 223
   multiple source-drain transistor in 223
   netlist file in 226
   node identification in 210
   node name in 227
   RC lines in 226
   similarity with DRC 183
   stamping in 210
   transistor size in 224
   wiring resistance in 225
   with corner stitched tiles 97

F

with raster   230
with trapezoids   230

Fairbairn, D.G.   87
false errors   184
fault coverage   277
fault grading   277
fault simulation   **276-278**
  concurrent   277
  parallel   277
fault simulator   11
faults
  correlation with defects   276
  detected   277
  undetectable   279
feedthrough   174
Fiduccia, C.M.   176
final value event   260
fixed menu
  in VTIlayout   105
floorplan
  of a datapath   159
  of PLA   149
floorplanning   8
Foley, J.D.   43
force
  on simulator inputs   246
Fowler, P.J.   225

G

gate
  boolean logic   7
  in circuit design   4
gate array   10, 168
gate simulator   7, 236
Gelatt, C.D.   173
global routing   168, **174**, see also routing

graphics package   44-65
  clipping in   47-50, 56
  color in   57-60
  coordinates in   **45-47**, 56-57
  driver in   45, 56
  input with   60-62
  mapping in   56
  primitives in   45
  scale in   55
  text in   55
  transformations in   50-54
  viewport in   47
  window in   55
Gray, J.P.   1
greedy
  in global route   174
  in initial placement   171
  in placement improvement   173
Gummel, H.K.   250

H

Hachtel, G.D.   153
Hamachi, G.T.   97, 104
hardware accelerator   248
hardware description language   6
Hartheimer, A.   120
Hartoog, M.R.   174
hierarchy   12-14
  advantages of   13
  and regularity   14
  creating in layout editor   101
  definition of   **12**
  display of   72
  example of   12
  flattening in DRC   187
  flattening in layout editor   101
  for faster searches   91
  in DRC and extract   228
  need to flatten   183
  searching in   90
  viewing level of   73, 98
high-impedance   242
horizon
  in edge algorithm   204
Horowitz, M.   225
Hsueh, M.Y.   111-112

Index

Huang, C.E. 225

I

icon 7, 98, 109
inertial delay 244
initial placement 170, see also min-cut
instance **12**
    in layout editor 82, 98
    in layout language 123
instance based layout generator 145
interactive command loop 70
interval arithmetic 257
intrinsic resistance 266
inverter
    CMOS layout of 4
iterating a design 12

J

Johannsen, D.L. 162, 167
Jouppi, N.P. 272

K

Karplus, K. 120
Kedem, G. 93
Kernighan, B.W. 170, 174
Kingsley, C. 112
Kirkpatrick, S. 173
Kliment, M. 145
Kozah, P. 250
Kuh, E.S. 174

L

lambda **19**, 184
Lanfri, A.R. 104
Lang, C.R. 1, 120
language based layout generator 143

latchup 4
Lattin, B. 1
Lauther, U. 188
layer
    as color 2
    contact 2
    diffusion 2
    DISCONNECT 228
    error 213
    metal 2
    NOCHECK 228
    number assignments for 21
    overglass 2
    polysilicon 2
    statement 21
    well 2
layout 8
layout editor 9, **81-108**
    as an extension to a plotter 82
    corner stitching in 95-97
    cut and paste in 101
    edit-in-place in 101
    editor state in 85
    essential features of 97-100
    filters in 85
    menus in 88
    pointing in 84
    problems with complex data structures 94-95
    quad tree in 93
    search tree in 92-95
    snapping in 84
    undo in 102
    unions in 90-92
    user interface of 102-107
layout file
    BNF for 18
    example of 24-25
layout generator 141-158
    conditional instances in 146
    instance based 145-148
    instance overlays in 147
    language based 143-145
layout language 119-135
    bounding box in 130
    building parameterized cells in 133-135
    cached cells in 121
    cell operations in 122
    connectors in 130-131
    contact in 126
    creating objects in 123

drawbacks of   136-137
embedded   120
example of   124-125, 128-129
extending   125, 131
transistor in   125
wire in   126
layout parser   17-38
layout language and   135
layout semantics
for plotter   32-37
for writer   38
Lee-Moore router   see maze router
left handed coordinate system   45
lexical analyzer   19
Lin, S.   170
local routing   168, see also routing
Locanthi, B.   120
logic design   7
logic optimizer   11
logic simulator   236, **242-245**

M

Mainsail   120
Martinez, A.   145
mask geometry   see layout
mask layers   1
mask tooling   229
Mayle, N.   137
Mayo, R.N.   97, 104
maze router   10, 175
McCormick, S.P.   225
McMullen, C.T.   153
McNamee, L.P.   225
Mead, C.   1, 39, 128
memory usage
of layout   37
menu   62
menu entry   62
metal layer   2
min-cut   **170-172**
min-cut partitioning   170-172
minterm   149

mixed-mode simulator   8, **272**
mode   86
acceptable   88
modeless editor   86
monte carlo   173
MOSIS design rules   185, 218
mouse   61
Mukherjee, A.   1, 184
multi-level simulation   272

N

n-channel transistor   3, 250
n-diffusion layer   2
n-well layer   2
Nahar, S.   173
Nance, S.   145
NAND gate
in CMOS   5
n-input layout for   142
net
in placement and routing   168
net compare   10
netlist
BNF for   226
definition of   **7**
file format for   226
in placement and routing   168
procedural generator for   136
static check of   269-272
Newman, W.M.   43, 49, 103
Niessen, C.   1
NMOS
depletion transistor in   3
inverter   251
node
charged   253
definition of   236
driven   253
forced   253
in transistor simulation   253
nonterminal   **18**
normalized device coordinates   46

Index

O

object-based data structure 32-37
    example of 34
observability 277
opaque colors 58
orientation code numbers 22
orthogonal layout 14, 98
Ousterhout, J.K. 57, 95, 97, 103-104, 113, 187, 272
overglass cut layer 2
overlay 147

P

p-channel transistor 3, 250
p-diffusion layer 2
p-well layer 2
painting
    in Caesar 103
pairwise interchange 163, 172
parallel fault simulation 277
parameterized cell 133, 141
    equivalence with instance substitution 165
Paroli, F. 16, 42, 68, 140, 182, 234
parser 17
    client of 26
    lookahead in 25
    operation of 29-31
    overkill with recursive descent 25
    semantics for 26
partition 7, *see also* min-cut
Pelavin, R. 176
pen plotter *see* vector plotter
phantom 98
physical screen coordinates 45
pixel **43**, 45
PLA
    as a state machine 156
    base and personalization of 152
    BNF for 154
    code file example 154
    code file for **153**
    core assembly of 150
    description of **149**
    don't-care term in 154
    electrical considerations in 157
    feedback in 156
    floorplan of **149**
    generator 11, **148-158**
    optimizer 11
    periphery of 157
placement 10, 168, **170-174**
    min-cut 170-172
        algorithm 171
    monte carlo methods for 173
    simulated annealing for 173
    weighted nets in 174
placement and routing 168, *see also* placement; routing
placement improvement 170
    cost function in 172
    in datapath compiler 163
plotter
    display in 72
    hardcopy output device 43
    necessary features of 72-74
    pen plotter in 75-77
    program 9, 69
    raster plotter in 77-78
point code 48
polygon merge 191-198
polysilicon layer 2
popup menu **64**
    in VTIlayout 105
procedural layout 9
procedural netlist generator 136
product term 149
property sheet
    in VTIlayout 107
pulldown structure 4
pullup structure 4

Q

query driven 70

R

raster plotter **43-44**
rectagon 101
recursive descent parser 25
Reed, J. 175
register transfer simulator 236
regularity of design 14
resistance
 of transistors 5
 of wires 5
resizing layout 199
 acute angle problems 200
river router 178
Rivest, R.L. 176
routing 10, **174-176**, see also
 channel route
 global route 174
 maze route 175
Rowson, J. 1, 87, 162, 167
RTL simulator 236
rubber banding **63**

S

SA0 see stuck-at faults
SA1 see stuck-at faults
Sahni, S. 173
Sangiovanni-Vincentelli, A. 153, 175, 236
Santomauro, M. 175
scan line **189**
Schaefer, T.J. 250, 266, 268
Scheffer, L.K. 228
schematic
 non-physical hierarchy in 114
schematic editor 7, **113-114**
Scott, W.S. 97, 104, 187
search tree 92
semantics 18
semantics module 34-37
 interface of 27-29
Shragowitz, E. 173
shrink see resizing layout
Shrobe, H. 137

signal values in simulation 237
silicon compiler 11, **159**, see also
 datapath compiler
 difficulty of definition of 159
simulated annealing **173**
 difficulties with 173
simulation 235-278
 breakpoints in 248
 changing the netlist in 247
 clock 246
 conditional break in 248
 controlling 245
 debugging in 264
 event queue in 248-249
 input
  clock 246
  forced 246
  propagation of 254
  stimulus language for 274
 mixed-mode 272
 multi-level 272
 netlist traversal in 247
 node in 253
 observing operation of 245
 output from 246
 single step in 248
 time quantum in 249
 trace output from 246
 transistor in 253
 transistor model in 252
 user interface for 248
slicing floorplan 177
snapping points 85
Soetarman, R. 228
sorting
 for edge files 190
 for pen plots 75
 limitations of 76
 of edge files 221
Soukup, J. 169
Sproull, R.F. 103
Sproull, R.S. 43, 49
stamping 210
standard cell 10, 168
 gates in 169
Starr, C. 145
state machine
 PLA implementation of 156
sticks editor 9

Index   291

stimulus   246
    language for   274
stipple   59
    for layers   2
structural description   7
stuck-at faults   **276**
Suetsugu, I.   249
Sussman, G.   137
Sutherland, I.E.   48, 85
switch simulator   236, **249-269**
symbolic layout   **108-113**
    compaction of   110
    compactor for   111
    disadvantages of   113
    editor   9, 108
    in a layout language   125
    transistor recognition in   112
    viewing in   110
symbolic layout editor   *see*
    symbolic layout
syntax   18
Szymanski, T.G.   210, 221

T

tablet   61
Taylor, G.S.   97, 104
technology file
    for DRC   218-220
    for extractor   224
    for layout language   126
    for plotter   74
    for transistor resistances   252
Terman, C.J.   230, 250, 257
terminal   **18**
test generator   11
test language   11, 274
    as simulator input   274
test vectors   11, 274
testing   **274-278**
    functional   275
    production   275
three-state simulator   242
tile   *see* corner stitched tiles

time
    and simulation accuracy   237
    source of delay   5
time wheel   **249**
timing analyzer   8, **272**
timing simulator   8, **236**, 249-269
token
    in parser   19
token scanner   19, **31-32**
trace output   246
track   **169**
track assignment
    in channel router   175
    in datapath compiler   163
tradeoff
    simulation speed versus
      accuracy   235, 244
transformations   **50-54**
    and orientation   22
transistor
    as a resistor   252
    as a switch   250
    depletion   3
    in netlist file   227
    in simulation   250
    layers for   3
    n-channel   3
    off   250
    on   250
    p-channel   3
    resistance of   251
    source and drain equivalent   250
    unknown   250
transistor simulation   7, 236
    algorithm for   **254-255**
    capacitance of a node in   259
    change to unknown in   259
    charge decay in   266
    charge sharing in   255
    combining resistances in   257
    delay calculation in   258
    driven stages in   256
    example of   261-264
    final value calculation in   258
    gate recognition in   269
    gate-like structures in   265
    resistor divider in   256
    stage in   253
    transistor usage in   266
    with three resistances for
      transistors   252

U

Trimberger, S.   1, 137

Ullman, J.D.   17
Ulrich, E.   249
unit delay simulation   240
unknown   **242**
    problems with   267
user coordinates   **47**
user interface   81, 102
user interface event   248

V

Van Dam, A.   43
Van Egmond, K.   274
Van Wyk, C.J.   210, 221
vanCleemput, W.M.   1
VDD   4
Vecchi, M.P.   173
vector plotter   **44**
viewport   47, 56
virtual grid compaction   111
virtual screen coordinates   45
VLSI Technology, Inc.   104
VSS   4
VTIlayout   104-107

W

Walker, B.   145, 162, 167
waveform   247
Weise, D.   137
Weste, N.   111-112
White, J.K.   236
Whitney, T.   228
Williams, J.D.   111
window
    CAD definition of   55
    graphics definition of   55

wire
    in layout editor   102
    in layout language   126
    in symbolic layout editor   110
    layers for   2
wrap number   **192**, 198

Y

Yoshimura, T.   174

Z

Zajicek, O.   272